T0185593

Resilience and Stability of Ecological and Social Systems

István Karsai • Thomas Schmickl • George Kampis

Resilience and Stability of Ecological and Social Systems

Springer

István Karsai
Department of Biological Sciences
East Tennessee State University
Johnson City, TN, USA

Thomas Schmickl
Department of Zoology
Karl-Franzens-Universitat
Graz, Austria

George Kampis
Eotvos University Budapest
Budapest, Hungary

ISBN 978-3-030-54562-8   ISBN 978-3-030-54560-4   (eBook)
https://doi.org/10.1007/978-3-030-54560-4

This Springer imprint is published by the registered company Springer Nature Switzerland AG
The registered company address is: Gewerbestrasse 11, 6330 Cham, Switzerland

We dedicate this book to our families, especially to our children, who may make this a better world.

Foreword

Climate change has been always affecting planet Earth, but what is different this time is that the human society became an important element of the causational network of these changes. Ecology and climate science are on the frontline of the research to decipher this network. Generating predictions of events are especially important for understanding and managing the processes of these complex systems. In this endeavor, modeling is an important skill set and tool to understand and predict. Unfortunately, however, the use of these tools is not widespread in biology because the training of biologists is generally different from the training of physicists or chemists, who are more accustomed to models and modeling. It seems that much of the existing excellent material (both books and articles) on either modeling or mathematical biology is not readily accessible to the students of biology. Mathematicians who are interested in modeling biological system are important to be inspired and communicated biology in ways what meaningful for them to produce relevant biological models or be able to collaborate with biologists.

Addressing the series of challenges and their combinations, humanity faces the increasing need for hierarchical thinking. Phenomena at the level of populations and societies, ecosystems and biomes are tightly linked and interact. We need both data and models that link up vertically the horizontal levels of organization, in order to better understand parts and wholes in nature and society.

This monograph written by a trio of long-term collaborators (István Karsai, George Kampis, and Thomas Schmickl) aims to promote the interdisciplinary field of mathematical biology. Their approach is focusing on accessibility. Instead of educating the reader in a textbook form, they show examples from their own research. A central theme is the resilience of ecological and social systems. Resilience is a central problem in climate change, sustainability, ecology, economy, and the social sciences. But the authors did not aim to write a textbook or a review of the field in a book format, but rather a carefully picked series of their research is turned into a monograph.

The first part of the book starts with a simple population growth model and leads the readers towards more and more complex problems. Stability of interacting species, turning fragmented habitats into a more stable state, or analyzing the effect of fire to ecosystems remains accessible to the readers because the models stay simple and the authors make these models accessible to the readers—encouraging them to do their own experiments.

The second half of the book is focusing on the stability and resilience of insect societies. These societies are the model system of ecological, economical, and societal systems. In some ways, they are simpler than ecological systems and human societies, and they are more accessible. These systems are also easy to manipulate with different perturbations; therefore, the theories on stability and resilience can be tested on them experimentally. These systems are also the fruits of evolution and they evolved several times independently. We can thus learn a lot from these systems on how we can help stabilizing system and managing resources. For example, the authors ask a simple question why bees do not hoard pollen, a resource which could clearly be hoarded and in fact commonly become scarce (for example, during rainy weather) in the colony. The lack of pollen will elicit cannibalism of larvae, slowing down the development of thousands of larvae and all of these events are happening several times a year. This all could be avoided if only the bees would store a bit more pollen. So why don't they do that?

Beyond these concrete examples and focused questions, the authors also contribute to the state of the art of the theory of stability research. Analyzing task allocation and material flow of different insect societies, they discovered a new type of regulatory network, which has not been discovered elsewhere above the cell level. The key finding that ensures stability is that multiple feedbacks go through a buffer node. While simple theory would claim that a simple negative feedback is able to provide stability, this would not work in real systems. Real systems need multiple feedback loops to ensure reliability and buffers to avoid unnecessary reactions. Current biological systems have been operating and evolving for a long time and they are not following the bare bone minimal theoretical structures, but they have redundancies. The authors show that not only these redundancies are important but also that they are easily accessible via theory.

In biology, we practically have no master equations to describe systems mathematically. We use mostly empirical functions, fitted data, or we use agent-based approach to model these systems. The authors show examples on how we can use master equations from physics to describe a biological system and how we can understand biology better by using these master equations developed originally to describe electric circuits.

Thus the monograph offers a wide spectrum of approaches, each adequate in its place, which together show the real complexity of biological systems that can nevertheless be understandable. This book is recommended for students and researchers interested in complex biological systems, from mathematicians to biologists and from engineers to social scientists.

Tihany, Hungary   Ferenc Jordán
January 2020

Preface

This book focuses on the way how mathematics helps us understand the ecology and how ecology will provide a new way of thinking in mathematics. While ecology has always been one of the most "mathematical" disciplines in the life sciences, an interesting schism has been evident in the literature. Experimental and field studies were carried out mostly by biologists, while theoretical work came from mathematicians. Cross-references between the two approaches were sporadic. Fortunately, this has been changed in the last decades and a plethora of work came out in this interdisciplinary field. So why do we need another book? What is so specific about this book?

Our goal here is to provide neither a textbook nor a review of the field. Our goal was to write a monograph on the subject on the basis of our own research in this field. Such an approach is essentially biased, opinionated, and may generate more questions than answers; all are properties of an ongoing scientific inquiry. We are very passionate about this field and we would like to radiate our enthusiasm for inspiring both mathematicians and biologists to venture into this field and to investigate. These days, it is not as hard as it seems. Cross-education is finally catching up and more and more people are educated in interdisciplinary fields such as data science, quantitative biology, applied mathematics, computational biology, and so on. However, there are other ways to get involved in this interdisciplinary field: collaboration. Currently, there exist several modeling platforms where people of differing background could work together using their own specific skillsets and produce something very new. These days everyone has the possibility and access to the free tools and tutorials to carry out these types of research.

Why do we emphasize modeling and simulations? What are these tools good for? One of the most important practical benefits by them is in the decrease of the complexity of the problem. In physics and chemistry, with a minimum ignorance, we can safely say that the units are the same. In biology, even two clones are not the same. The enormous number of variables and their even larger number of interactions make biological problems highly intractable. The main step in modeling is the decreasing of the number and complexity of variables to the minimum.

This step translates the theoretically intractable original biological problem into a manageable abstract system, where all variables and their interactions are known. Also, transferring a biological system into quantitative models removes ambiguities of terms and makes assumptions explicit and clear. Generally, this step sharpens the goals and the questions and makes it impossible to sweep things under the carpet, so to speak. This is the hardest and essential step that requires both scientific (biological) and mathematical thinking. Choosing key variables and relationships (functions) and neglecting others lead obviously to a simplification. Is this simplification a problem? Yes and no. Yes, because we are actually not studying the biological phenomenon itself, but only its simplified mockup. No, because we can understand this mockup much easier than the real system. We can get an insight on how the original system works. With different models, we gain different insights. This is the main reason that there is no definitive model or modeling technique for a given biological problem. This is also the reason why in this book we stress the importance of different modeling techniques and an iterative approach of modeling.

The title of the book is "Resilience and Stability of Ecological and Social Systems," which covers very well the content of the book. However, our scientific work in the last 2 decades was not focused on just studying stability itself. This was a meta-result of a series of focused researches we had been conducting. Originally, we simply wanted to understand how some ecological and social systems work. Our goal was to describe a system mathematically and try to understand its basic properties. As we studied several of these specific systems, a common theme emerged. These systems all are self-organized and a network of feedback loops ensures their resilience and stability. For this book, we picked those works from our studies where we have something new to say about a specific biological system, but they can also be used as good examples of the general theme. We also have as a main goal that this book should be accessible for students of nature (even for those whose background is sporadic in modeling). To achieve this, we start the book with a simple case of population growth, then we continue with different types of population interactions and finally we reach the more complex topics of the functioning of insect societies. Without going into textbook level details, we believe the structure of the book makes easy to access it for both mathematicians and biologists.

To enhance this accessibility of our studies, we refer to our original papers and we make all computational models available.[1] Thus the reader has the option to use our models for her/his own investigations. When we describe a system in this book, we focus only on the core of the model and for all further details, such as those of sub-models, parametrization, and so on, we refer to the original technical papers. We also provide a short appendix on modeling techniques. We provide a series of models that will run on free or open-source modeling platforms. These platforms

---

[1] https://sites.google.com/site/springerbook2020/home.

have detailed tutorials and they are very accessible even for undergraduate students. We want you to experiment, rewrite our models, have new ideas, find shortcomings in our work, and have a never-ending adventure in the field of science.

Johnson City, TN, USA   István Karsai

Graz, Austria   Thomas Schmickl

Budapest, Hungary   George Kampis
Kaiserslautern, Germany

Acknowledgements

The writing of this book was partially supported by a sabbatical leave from ETSU and a Fulbright Scholar Grant for István Karsai.

Contents

Chapter 1
Understanding Ecosystem Stability and Resilience Through Mathematical Modeling

Abstract This introductory chapter of the book has two main goals. First, we outline why a stronger connection between Biology and Mathematics (especially modeling) could provide new insights into current biological problems. Second, we map out the central theme of this monograph and the logical connections among the following chapters with each other. We believe that one of the central and most important issues of how society can survive in our fast-changing world is to understand and learn how biological systems are able to stay stable and resilient against many perturbations. Besides providing a "big picture," an overview of some of the important aspects of ecosystem stability, we also present tools and approaches that make it accessible to the reader. We are taking examples mostly from our own research, make our models accessible, encourage further experimentation and involvement in this field. Ecosystems can teach us how resilient mechanisms can operate in a decentralized and dynamic way, driven by self-organization, which is achieved through several interacting feedback loops. Knowledge derived from biological systems has already inspired technology and arts. In our opinion, the same kind of bio-inspiration will soon influence politics, governance, economics, and other important aspects of human society as well.

1.1 The Interface Between Mathematics and Biology

Mathematics do not have the same long-standing relation to the realm of life sciences that it does to physical sciences and to engineering (May 2004). This is not due to a lack of interest or recognition of the relevance of the issue. For example, D'Arcy Thompson famously began his book (Thompson 1917) "On Growth and Form" in 1917 by quoting Immanuel Kant: "[...] chemistry was a science, but not a Science[1]... for that the criterion of true Science lay in its relation to mathematics." Thompson then went on to explain how Chemistry was elevated to the level of

---

[1]German nouns are always capitalized. So the citation does not work directly in the present form—but this is how it became famous. . . .

© Springer Nature Switzerland AG 2020

I. Karsai et al., *Resilience and Stability of Ecological and Social Systems*,
https://doi.org/10.1007/978-3-030-54560-4_1

1

"Science," whereas Biology had not reached that level yet, at least in the first half of the twentieth century. It seems that the fascinating diversity of life promoted more of pursuing empirical discoveries than did synthetic or theoretical studies. For example, Darwin established one of the main theories in Biology, yet he never formalized it mathematically. He recognized this fact and regretted it: "I have deeply regretted that I did not proceed far enough at least to understand something of the great leading principles of mathematics; for men thus endowed seem to have an extra sense" (Darwin 1887). These days, Mathematics enters every stage of Biology: in designing an experiment, seeking response patterns, and in the search for the underlying mechanisms. Biologists have applied an interdisciplinary approach, both in research and education (Karsai et al. 2011) and have been developing joint simulation platforms, where mathematicians and biologists can work together (Karsai and Kampis 2010).

Can any effect of a collaboration with Biology be observed on Mathematics? The epistemological origins and approaches of Biology and Mathematics are different, and thus their approach to gain knowledge also shows considerable differences. Nonetheless, the interface between Biology and Mathematics has initiated and fostered several new mathematical areas (Sturmfels 2005). For example, early population geneticists (Fisher, Pearson, Wright, and Haldane) had developed a mathematical formulation that Darwin was unable to assemble. They not only provided a mathematical foundation for the theory of evolution by natural selection, but in the process also produced several major advances in statistics (Shou et al. 2015). Another example from the past is Robert Brown, a botanist by training, who has discovered what is now called Brownian motion, via observing the movement of pollen grains in water. Later, the mathematical description of such motion became central to probability theory (Levin 1992). If we are looking at the big picture and not just at interesting examples, we will discover three major inspirations that Mathematics took from Biology (Reed 2015).

First, the theory of evolution and genetics invigorated the fields of probability, statistics, and stochastic processes (Wright 1984; Karlin 1986). Second, the Hodgkin–Huxley equations (Hodgkin and Huxley 1952) and Turing's work on morphogenesis (Turing 1952) inspired research in pattern formation (including traveling waves) and reaction–diffusion equations (Fife 1979; FitzHugh 1969). Third, sequencing genomes created new inquires in probability, statistics, and combinatorics (Karlin and Altschul 1990).

Biology will progressively stimulate the creation of qualitatively new fields of mathematics now as well as in the future. This view stems from the vast organizational complexity of living beings. In Biology, emergent properties, which are often called ensemble properties in Mathematics, are present at each level of organization. These properties emerge from interactions of heterogeneous biological units at a given level and also at lower and higher levels of organization. To understand the complexity of this biological organization, new mathematics will be required that can cope with both, with the ensemble properties and with the heterogeneity of the biological units that compose ensembles at every level (Cohen 2004). For example, the modeling of the transport of materials in axons led to new theorems in partial differential equations (Friedman and Hu 2007). The theory of biochemical reactions stimulated new theorems in dynamical systems (Anderson 2011) and in queueing theory (Mather et al. 2011).

Our goal here is neither to provide an exhaustive list on where and how biology has promoted mathematical research, nor will we try to list up how Biology was enhanced by Mathematics, except in some exemplary cases. For such an overview, we refer to the excellent essay of Bellomo and Carbonaro (2011) or Bellomo et al. (2017). Cohen (2004) listed cellular networks, brain and behavior, evolution, ecology and environment as major biological challenges which might have the potential to stimulate major innovations in Mathematics, and, in turn, to also benefit from those innovations. In this book we will focus on two of these areas especially, how interactions among organisms and their interaction with the physical or social environment can stabilize their ecosystems and why emergent (i.e., ensemble) properties of these systems are important in the process.

There is some difference of how mathematicians and biologists perceive the term *stability*. Grimm and Wissel (1997) inventoried 167 definitions used in the literature and found 70 different stability concepts. Without getting into too much details we will outline here how we handle this and similar terms in this book. In Mathematics, stability theory generally addresses the stability of solutions of differential equations and of trajectories of dynamical systems under small perturbations of initial conditions. This is sometimes also called Lyapunov stability (Parks 1992). This mathematical approach is very important for model-making and testing. This approach is used when the robustness of the models is tested in terms of how sensitive the model is to different starting values of the parameters. We tested our more complicated models with this mathematical procedure and we call the process sensitivity analysis to avoid confusion with the more biological terms of stability.

In Biology, namely, an ecosystem is said to possess ecological stability if it is capable of returning to its equilibrium state after a perturbation. This capability is known as resilience. A biological system is also called "stable" when it does not experience unexpected large changes in its characteristics across time (Levin et al. 2012). When we talk about stability or resilience, we mean this along the given biological definitions of the terms. It is important to have further descriptors on these terms, because the approach of biologists is typically less rigorous than required by the mathematical definition. Therefore, we have to specify the type and the entity of the stability in question. It is possible for a biological system to be stable concerning some of its properties and unstable concerning others. For example, a plant community might, in response to a drought, conserve biomass but loose its biodiversity (May and McLean 2007). It is also noteworthy that biologists may call a system to be stable even if it is not at equilibrium. For example, predator–prey systems could be described as persistent and resilient, yet they are not constant but change in wide ranges but within specific bounds. Also, many ecosystems that are held "stable," actually oscillate around an unstable fixed point, without ever staying at equilibrium.

1.2   The Significance of Environmental and Ecological Change: Stability and Resilience

As mentioned above, ecosystems are not necessarily stable in a general mathematical sense, but show rather dynamic resilient behavior against internal and external disturbances in the long term. Biology often stresses the effects on both kinds of perturbations. At the individual and population levels these effects are commonly grouped into two categories: (a) *biotic factors*, which originate from other organisms, for example, via competition, predation, parasitism, etc. and (b) *abiotic factors*, which originate from non-living sources, such as weather, seasonal changes, climate changes, geological processes, or the consequences of natural catastrophes, which are often just severe instances of change due to one of the abiotic factors mentioned before. While this line of separation is part of a classic perspective in biology and ecology, it is not necessarily useful when trying to understand, or even predict, the emerging population dynamics and the ecosystems' stability features. Thus, when looking at the stability and resilience features of ecosystems, we prefer to categorize the effects of dynamic factors along a different line of separation, into again two different categories:

- *Extrinsic factors*: These factors may be biotic or abiotic, their important characteristics being that these factors affect the focal ecosystem without regulation in any significant way from within the system itself. One could see these factors as ones that are "shaking" the system from the outside, which can perturb the system and move it away from its equilibria. But these external factors are only relevant if they change over time, as a constant steady influence will not challenge the existing equilibria we already observe. Dynamic external influences can appear in two basic forms: Changes over time that are predictable, for example, tidal, diurnal, lunar, or seasonal cycles. We usually incorporate such influences as a trigonometrical (such as a sine) function in the models. Whenever these extrinsic factors do not follow a predictable pattern, we can interpret them as a form of "external noise." Practically, it does not make any difference whether these factors are really purely random or if they are just originating from another (external) system of which we do not have a clear model. Therefore, we usually model such cases both by random extrinsic factors in our mathematical description. Still, even though such factors are considered random, statistics may provide quite specific information about them, for example, by observing their upper and lower bound, maybe even the shape and other characteristics of the frequency distribution of states.
- *Intrinsic factors*: These factors affect the focal ecosystem's equilibria and originate from specific interactions between the parts of the system itself. For instance, very often there are specific forms of oscillations that we can observe in ecosystems, most prominently in predator–prey systems, but also in host–parasite systems, a realm into which also all epidemic dynamics fall. We usually find two key ingredients by analyzing the mechanisms producing

these oscillations: It is either feedback loops built from reciprocal interactions of system components and/or time delays within such interactions. Both of these are among the most essential components of biological functioning. A feedback loop basically means that one system component (e.g., one organism species) affects the dynamics of another component (e.g., another species) and the other species affects in turn back onto the population dynamics of the first species. In Biology, longer and more complex feedback loops, constituted by many components, are also very common and therefore the interactions between the components are rarely direct, but are rather closed circles of cascading effects that ultimately feed back to the first element of the loop. Feedback loops can come in two different forms: (1) positive feedback loops, which can drive systems towards increasing oscillations and extreme system states and (2) negative feedback loops, which stabilize systems and drive them towards non-extreme equilibria states. Usually in ecosystems both types of loops will coexist and balance each other out, but time delays or occasional perturbations from extrinsic factors can lead to oscillatory behavior of such systems on the long term. When multiple feedback loops occur, and especially if these feedback loops operate on different time scales, more complex dynamics may occur. Some of these dynamics are of such a high complexity that they look chaotic, and these systems may require complexity analyses for deciphering the mechanistic explanation of the emergence of these patterns.

Biological systems have a strong resilience against perturbations and they can adapt to these perturbations to some extent. Having a point attractor as an equilibrium and keeping the state of the biological system close to this point might not result in the most stable and resilient system behavior. It seems that oscillations and dynamic changes are system-inherent features in ecosystems, and natural selection has not "nailed down" population dynamics rigorously. We may conclude from this observation that natural selection favored processes that create dynamic and flexible resilience features over ones that show rigid and robust stability features. On an evolutionary relevant timescale, frequent dynamic change of population levels might positively affect the adaptation process of the focal species, as oscillations of populations lead to dynamic changes of gene pool sizes and to recurrent changes of selection pressures. Both of these factors are known to have an effect on the dynamics of evolutionary adaptation, most prominently known as in its extreme form, the so-called bottleneck effect, which is known to accelerate the proportion of mutants in the population.

On an ecologically relevant time scale, dynamic equilibria might allow species to adapt to changes in resource abundance in an adaptive way. On shorter time scales, observed oscillations can be also tightly connected to the life-cycle regime of specific organisms, often adapted to seasonal changes in resource abundance.

Ecosystems are adapting their configuration in a dynamic way so that the species they contain minimize their overlap in their required resource spectra, known as

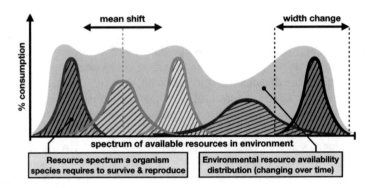

Fig. 1.1 Schema of an exemplary distribution of the resource spectra required by several species that share a habitat. Each species requires a given amount of resource (area below the curve) to survive and reproduce. The resource usage is dynamic, especially when the overall resources availability is dynamic as well. Every change in one population's resource usage affects the overall resource availability, and is, in consequence, ultimately affecting the population dynamics of all other competing species that depend on the same part of the resource spectrum

"Gause's law," or competitive exclusion principle,[2] (Hardin 1960). Adhering to this principle not only makes overall resource usage more efficient, but also it complexifies the ecosystem over time, and this in turn influences the overall stability of the system. In most ecosystems many species still coexist side-by-side with a significant overlap of their resource demands in the spectrum. The overall system is likely seeking a configuration of overall species-resource-distribution that represents a sort of "lowest energy state" equilibrium, meaning, minimizing the overlaps but still exploiting as much of the available resources as possible (Fig. 1.1).

To adapt in such a configuration is a complex process, especially in changing environments. Basically, such an adaptation process is a form of an iterative evolutionary search process to find a stable common configuration across all involved species within a vast high-dimensional search space. This search space is defined by all possible combination of physiological and morphological parameters that may be "tuned" (i.e., adapted) through natural selection of each species, as well by the plethora of possible ecological interactions between the species that share common resources in the same habitat. Thus, these ubiquitous multi-level adaptations in their totality resemble a process where all species are searching quasi-optima in a complex, thus non-trivial, search space—and do that in a massively parallel fashion. As Hills et al. (2015) put it nicely:

Search is a ubiquitous property of life.

Various disciplines of science, for example, Mathematics/optimization theory (Lee and Geem 2005), Computer Science/machine learning (Larrañaga et al. 2006),

---

[2]The competitive exclusion principle is classically attributed to Georgii Gause although he actually never formulated it.

psychologists and biologists studying human and animal cognition (Olton and Schlosberg 1978), as well as economics (Peli and Nooteboom 1999), have all studied the efficiency and effectiveness of such dynamic search and adaptation processes. Searching in difficult search spaces depends on having a good ratio of exploration to exploitation strategies, which ultimately also yields the requirement of having a good mixture of generalist and specialist species in the ecosystem.

The ratio of exploitation to exploration is a key feature of adaptation. Any form of search or optimization process has to develop over time by exploiting a sort of gradient, which is driving the system towards better system states. The benefit of exploitation is that it improves the system's state over time, and the drawback is that the system can easily be trapped in a local optimum. In contrast to exploitation, exploration characterizes system changes that do not strictly follow a gradient, e.g., random behavior or larger "jumps" concerning system states. These acts can drive the system out of local optima and will allow it to progress towards a global optimum if the ratio of exploitation to exploration is well-balanced. If this ratio is dynamic, optimization processes will often perform even better, thus many modern heuristics, namely stochastic optimization methods, employ such adaptive exploitation–exploration ratios.

Oscillations commonly observed in population dynamics may very well stem from these search heuristics. When population size is large, many genetic variants will exist in parallel and their fitness will be tested in the given environment, thus exploration will be high in this phase. However, at high population size a large fraction of the available resources will be used up by the organisms. In consequence, selection pressures will increase, and intraspecific competition will shift the exploitation-to-exploration ratio towards more exploitation, while population sizes will start to decline. This process resembles key aspects of heuristic optimization algorithms, for example, "simulated annealing" (Kirkpatrick et al. 1983), which is based on a cyclic pattern of repetitive ups and downs in the exploitation-to-exploration ratio. A similar process may also improve the local adaptation of organisms to their environments by allowing the population to oscillate by some extent around an equilibrium point, instead of keeping the population on a fixed level all the time. The overall effect might be a quicker rate of adaptation, thus a better coping with environmental change, thus it will lead ultimately to a more stable ecosystem concerning long-term survival.

The evolutionary search for a quasi-optimal solution requires compromises and trade-offs. A related "no free lunch" theorem was first coined in computer science for search and optimization algorithms (Whitley and Watson 2005) but it also can be applied to natural mechanisms of adaptation. Traits and mechanisms that perform well in some circumstances will perform worse in other circumstances. Laying more eggs is commonly coming along with a trade-off of shorter fecund longevity. The more a specialized mechanism can exploit a specific set of circumstances, the less suitable this mechanism will usually be in very different circumstances (for example, for exploration).

Trade-offs also emerge and control how organisms interact with each other and their environment. Some members of communities act as generalists, while others act rather as specialists. This is well-known, concerning the resource demands of species in ecosystems, organizations participating in markets in economy, as well as in the division of labor in animal societies. In general, specialists allow for a rapid and effective exploitation of resources in situations, where the latter are plentifully available. Generalists bring stability to the system by distributing resource consumption widely, over the whole spectrum. Specialists usually have fewer ecological interactions due to their specialization, but these interactions could be crucial for other species. For example, an extinction of a pollinator specialist may result in the extinction of a whole plant species. Complex ecosystems with a high number of specialists, like the ones found in rainforests, can be very vulnerable and their stability can be lost when some of the key specialists go extinct. Generalists may, by contrast, have many ecological interactions with other organism species. This makes generalists more likely to be participants in various feedback loops that may stabilize the overall system. So, basically, the varying degrees in specialization of natural organisms can be seen as an embodied form manifesting a specific exploitation-exploration ratio.

Due to dynamic extrinsic and intrinsic factors and time delays, the population dynamics of organisms can be described as an oscillation around a non-zero equilibrium point. All populations are governed by negative feedback loops that limit not only their growth, but also keep these oscillations within certain bounds. These feedbacks arise mainly from inter- and intraspecific competition, as well as from food chain interactions, which are all due to the limitations of their shared resources. These shared resources might be organisms themselves—regulated by a similar set of fundamental negative feedback loops. This chain of causation continues further down to the basic resources provided by the "producers" within this ecosystem. These organisms' growth is limited by basic principles of physics, for example, by the limitations coming from the conservation laws that govern energy and mass. Biological feedback loops are also strongly connected to physical and chemical loops, such as the nitrogen cycle, the CO_2 cycle, or the global heat regulation of the globe, via albedo provided by the plant (such as foliage) coverage. Thus, ultimately, these basic physical principles do not only limit the growth of all occupants of ecosystems, but they also stabilize the oscillations of the population dynamics of the organisms.

1.3 System Changes that Can Lead to a Loss of Stability and Resilience in Ecosystems

Due to extreme oscillations or perturbations, the population can reach another extreme stable point when population size becomes zero. This is a point of no return in population dynamics ("privileged zero" property), at least if the given species' population in the focal ecosystem cannot be bootstrapped again by immigrating organisms from other habitats. If this is not possible, the species becomes extinct in the given habitat. Such an event leads to a permanent structural change of the focal ecosystem, as one key actor is permanently lost, potentially affecting the stability and resilience capacities of the ecosystem in the future.

Today, we are increasingly facing a period of rapid severe ecosystem change. More and more studies show that ecosystems get thinner and thinner across various groups of organisms, such as insects (Hallmann et al. 2017) or vertebrates (Ceballos et al. 2017) tend to disappear all around the planet. We say that an ecosystem becomes "thinner" if it has either lost a significant number of species in its composition, or if the existing species have lost the strength of their interactions with other species due to a significant decline in the population size. Studies are reporting alarming figures from up to 70+% biomass loss and 70+% loss of the occurrence of organisms in their original habitats (Hallmann et al. 2017; Ceballos et al. 2017). As an example, about a decade ago, the general public has received news about the first alarming signs of massive collapses of the honeybee population. Scientific studies have started to report the emergence of a novel phenomenon, the "Colony Collapse Disorder" or, in short, "CCD" (Ellis et al. 2010). Today we know that CCD is just the tip of an iceberg (Fig. 1.2). In fact, we are neither facing a "honeybee colony collapse" only, nor are we facing an "insect apocalypse" only, a term first coined by a prominent New York Times magazine article (December 2nd, 2018). Instead, data indicates that we might be actually facing the start of a new mass extinction period (Ceballos et al. 2017).

The degree of this decline is difficult to measure, especially on a global level. Measuring physical environmental properties like the average temperature or the concentration of CO_2 is a rather straightforward task, albeit involving specific

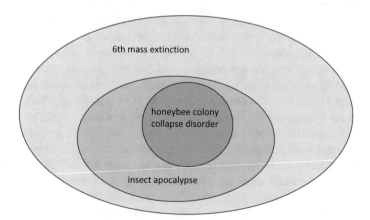

Fig. 1.2 In the current ecological crisis the public became first aware of the honeybee colony collapses around 2007, before articles about an "insect apocalypse" started to spread occasionally. The broader public got aware of the possible beginning of the sixth mass extinction only in mid-2019, when the IPBES (2019) summary report was published

machinery. In contrast, measuring the biological well-being of ecosystems is way trickier. On the one hand, this requires the difficult counting of organisms across habitats, what is a tedious task by itself. On the other hand, it is impossible to break these measurements down into a few numbers to qualify the current state of an ecosystem, as not every species has the same significance for the stability of an ecosystem. Thus, to assess the well-being of an ecosystem means to understand the composition and the dynamics within it and to have the ability to interpret these values. Such an interpretation can be provided by mathematical models that predict the future stability and the resilience of a given system. There can be multiple stressors acting together in affecting organisms within ecosystems synergistically (McGill et al. 2015), potentially adding up or even multiplying their combined effects. While many believe that climate change is a main factor here, global warming is just one out of the many factors that can affect ecosystems negatively: Pollution of habitats, over-fragmentation of habitats, invasive species through globalized traffic and logistic networks, over-exploitation of habitats by harvesting populations or by depleting their resources, loss of habitable land by forest fires, erosion, desertification and construction, as well as rapid transformation of the physical habitat properties are but just a few of the anthropogenic stressors that may have triggered the current ecosystem decline.

While these stressors may trigger ecosystem decline, the consecutive dynamics of the collapse are determined by the internal structure of the systems, and cascades of extinctions and co-extinctions might occur due to the ecological relationships and dependencies among organisms (Dunn et al. 2009). Every natural ecosystem is a complex network of behavioral interactions and substance flows that carry energy and nutrients from one individual to another. The stability of the overall system depends mainly on its intrinsic structure emerging from these local and individual interactions. Most prominently, negative feedback loops govern the stability and resilience of an ecosystem (Bennett et al. 2005). Such feedback loops arise from processes of saturation or depletion, and ultimately their causation derives from the basic conservation laws of physics. And here is the core of the problem: Whenever a species shrinks in population size, all feedback loops (including stabilizing ones), in which the given species participates, will become weaker. At the extreme, when a species goes fully extinct within an ecosystem, all feedback loops it was participating in will cease to exist, removing all their stabilizing effects from the ecosystem. In consequence, this loss of stabilizing feedback loops can then lead to more ecosystem disruption, in the form of sudden oscillations and other sudden changes of conditions within the remaining part of the ecosystem. This will in turn stress the remaining other species or even driving them towards extinction. This way, the whole dynamics can turn into a vicious cycle, in which ecosystems can very suddenly collapse and loose many of their species within a short period of time. Diversity of species composition can, on the other hand, bring stabilizing feedback loops into action, thus making them stable against the disruptive (often negative) effects (Downing et al. 2012).

In conclusion, a sort of vicious cycle can be triggered within all ecosystems: A lack of stability and resilience can favor detrimental structural changes, like diversity

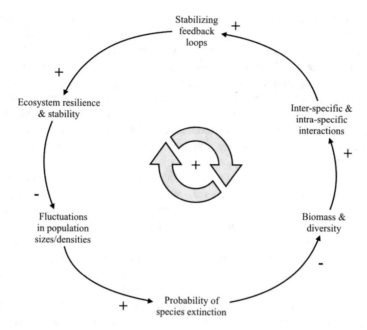

Fig. 1.3 Conceptual drawing of the escalating feedback loop that may drive an ecosystem towards collapsing due to the diversity loss and the resulting effect on stability and vice versa

loss and biomass loss, which may in turn again affect the ecosystem's stability and resilience properties in a detrimental way. There are compensating processes at work in nature that might keep this cycle of decay ineffective for long time spans, for example, the re-immigration of species from other habitats in metapopulation systems and the adaptation of the remaining species to the novel situation. However, this vicious cycle, as it is schematically depicted in Fig. 1.3, may take over the system dynamics when a certain tipping point is exceeded, for example, after a strong external trigger event or a significant alteration of the living conditions.

1.4 Understanding Ecological Stability and Resilience: Our Modeling Approach

What can we do to prevent or to slow down this kind of ecosystem decline? Obviously, the first step is to thoroughly study our ecosystems and to gather as much data and insights about their internal processes as possible. A second step is to consolidate this knowledge into a structured form that allows us to understand the observed dynamics, but also allows us to predict future dynamics in response to structural changes in a given ecosystem.

Mathematical models are important tools to generate such a form of understanding in a way that knowledge from various sources can be consolidated and also checked for consistency. In this book we present an overview of the work we have conducted in the past decades to generate exactly these types of models. We did neither restrict ourselves to a specific modeling technique nor to a specific mathematical model implementation type (Appendix A). In contrast, we built top-down models as well as bottom-up models including multi-agent models, cellular automata models, stock and flow models, and ODE models. Our focus was to investigate how interactions within a system and between systems can affect their stability and resilience.

Besides the general theoretical goal of a better understanding of the prerequisites for stability and resilience in general, our models cover important aspects of biological populations (growth, structure, interactions, and external effects). For example, we investigate the effect of forest fires for both trees and forest animals. The latter is frequently ignored in studies of similar issues (Chap. 4). This topic is highly relevant for investigating the significance of burning down rain forests for agricultural development, or for investigating the feasibility of "prescribed fires" in order to conduct proactive fire control. Such fire events may seem to be localized in their effects as tree regeneration in the affected area might commence soon thereafter. However, these events also have global-level effects, as the ecological health and the overall size of forests affect the global climate (Flannigan et al. 2000). Besides the practical recommendations we can provide with our model, we also investigate the validity of existing theories and thus promote the understanding of forest fires concerning their effects on the stability of ecosystems.

We also investigate how fragmented habitats and the migration within sub-habitats can affect population dynamics (Chap. 3), a topic highly relevant in our ever-more fragmented biosphere. The existing ecological literature condemns habitat fragmentation to a large extent. However, nature itself creates highly fragmented habitats on a frequent basis: Every tree is branching, and every branch has many sub-branches and sub-sub-branches, and so on. Ultimately, these branches lead to many leaves, which also have often fragmented subparts as an upper and a lower side. Thus, looking at a simple forest habitat formed by many trees and bushes, one can see that nature is producing a fragmented habitat of many layers nested into each other. Similar observations can be made for almost all other habitats that develop even under fully natural conditions. Studies of fragmented predator–prey systems (Janssen et al. 1997) have shown, for instance, that a certain degree of multi-level fragmentation even supports species coexistence and thus ecosystem stability. The important question is, what we can do with habitats that get fragmented by human interventions beyond what we can observe naturally? A specific fragmentation model we have developed (Chap. 3) suggests the benefit of keeping several large enough habitat fragments in a healthy status that are connected with each other through corridors allowing a minimal migration between these fragments. Under such conditions the whole ecosystem can be stable globally. Fragmentation paired with migration corridors could even be more beneficial for the most vulnerable to predators than the original untouched big habitat. Is this counter-intuitive? Yes, it is. It is thus important to keep in mind that what sounds logical is not necessary a scientific fact. Modeling is an important tool to test such "what-if" scenarios.

Model building and scientific thinking are, in general, simplification processes. We try to focus on important things and neglect the less important ones. But how do we decide what is important for a given study? There is always the danger that we oversimplify our view and then build a "toy model" in consequence, which cannot provide useful predictions and understanding of the real system it should depict. For example, most models simplify populations to a homogenous mass, but natural populations are structured by age, sex, and other significant life history factors. Migration can be associated with age (Niemelä et al. 2000). Sexual population structure can also be an important driver of population growth and for migratory patterns, especially when sex ratios are skewed (Janzen 1994). The main issue we put forward here is not about having more parameters in the model in order to make a model more realistic. In contrast, we argue that implementing specific key life history parameters will allow for a completely different understanding about population growth and stability then we had before (Chap. 2). Also, here the theoretical understanding of stability itself is very important, but our models have strong practical relevance in species that show population declines (Allee et al. 1949) and in species in which temperature affects the sex determination, especially as temperatures are predicted to raise significantly in future (Kallimanis 2010).

We also emphasize that population interactions are key to understand the stability and resilience of ecosystems. Elementary population interactions such as competition or predation have been the center of population ecology research for a long time. However, at the community level, these population-level interactions form a complex network and they actually interact (affect) each other. We model intraspecific competition among organisms as a basic core dynamic of ecology (Chap. 2). Intraspecific competition does not only regulate population density and has, as a consequence, an enormous stabilizing effect on ecosystems, but it is also a main driving force in evolution. In sexually reproducing species, conspecifics compete for mating partners (this is called sexual selection) and in general all individuals within a species compete for an almost identical resource spectrum (i.e., natural selection).

The other form of competition is interspecific competition, which is important for ecosystem stability, as it creates additional negative feedback loops within the ecosystems. Negative feedback loops are known to increase robustness and stability (Bennett et al. 2005), thus the understanding of these aspects is important for predicting ecosystem behavior. Predation in its general form (i.e., the consumption of living beings, including grazing) is another population interaction which is covered by a large body of literature. However, how predation "interacts" with intra- and interspecific competition within an ecosystem is still elusive.

Generally, the interaction between trophic layers is imposing another negative feedback loop, thus having a further stabilizing effect on the system. Our model tests the hypothesis that predators are stabilizing their ecosystems, because predators are believed to decrease the level of competition between the consumed prey populations (Chap. 2). However, our results show the opposite. Predators are introducing oscillations in the systems. In a spatial model of predator–prey systems in fragmented habitats (Chap. 3), we show that such numerical oscillations can produce peaks in the population dynamics that spatially spread across the habitat in wave-like manners. The interactions between predators and prey determine the local densities of the affected populations, however, these local densities might be important for the spread of other effects, like parasites, diseases, or forest fires (Chap. 4). Moreover, we show that these oscillations can be so severe that one or several co-dependent populations go extinct.

Nature has not only evolved competition and food chains, which negatively affect at least one of the partner populations. In addition, interactions with positive influences between organisms have also evolved, such as cooperation, symbiosis, and social life. These are important aspects that determine population dynamics and again ecosystem stability. We have extensively studied cooperative behavior in the resource allocation of insect societies (Chaps. 5–7), because these societies are excellent model systems for both ecological and societal (including economic) systems. Social insects are easy to study and manipulate experimentally and so we can investigate how these societies regulate resource gathering, resource allocation within their society (colony) and also their working activities, in a very resilient and sustainable way.

These regulation mechanisms have been "carved out" from specific empirical studies and from a set of detailed specific models (Chaps. 5 and 6). We have also developed a core model of mechanisms which we call "common stomach regulation" mechanism (Chap. 7). The beauty of this regulation mechanism is that it allows the insect societies to harvest their environment in a sustainable way. Nothing is harvested that cannot be used in the colony within a short time frame, while the workers, at the same time, maximize the efficiency of their ecosystem service. The mechanisms employed by these animals are also very simple and act only locally, without the need for a global overview of the system's configuration.

This generalized model can point towards a sustainable resource utilization strategy, where not maximizing the profit but maximizing stability and reliability are the most important aspects of operations. It is also noteworthy that such a system neither requires planning for the output nor requires to inflate the demand, but is in fact self-organizing around a balance of a stable ratio of supply and demand. Such a "common stomach" mechanism can also be employed in other domains, and at various size and time scales. Thus, it can allow us to draw bio-inspiration from animal societies for designing future bio-inspired technology or for future policies of resource usage and allocation. This way, it may further the growing of our human society towards policies and technologies that respect the natural limits of growth, as we know them to exist, but which we have chosen to neglect for a long time (Meadows et al. 1972; Turner 2008). We hope that this book gives a good overview of the rich set of aspects that has to be considered in order to understand the dynamics of ecosystems. This is essential for predicting the future functioning of ecological and social systems. Only by developing such an understanding of ecology can we work efficiently and effectively on stabilizing and restoring broken ecosystems, of which we see many examples today. Ecosystems can teach us how resilient mechanisms operate in a decentralized and dynamic way, driven by self-organization achieved through interacting balancing feedback loops. These systems are complex in their internal structure, and it took natural selection ages to challenge, to adapt, and to evolve them throughout long times in the past. Systems that we can observe today have survivors that have successfully passed this test of time and today show an impressive level of resilience and stability. They are able to inspire us to take them as templates to construct similarly robust systems in other domains. Mathematical models are the best tools we can use to encapsulate this knowledge in a general and understandable way. Such generalized biology-derived knowledge has often inspired technology and arts. In our opinion, bio-inspiration should not only inspire these domains of societal progress, but we can foresee that bio-inspired politics, bio-inspired governance, and bio-inspired regulation should become the next Big Thing.

References

Allee WC, Emerson AE, Park O, Park T, Schmidt KP (1949) Principles of animal ecology. Saunders, Philadelphia

Anderson D (2011) A proof of the global attractor conjecture in the single linkage class case, SIAM J Appl Math 71(4):1487–1508

Bellomo N, Carbonaro B (2011) Toward a mathematical theory of living systems focusing on developmental biology and evolution: a review and perspectives. Phys Life Rev 8:1–18

Bellomo N, Bellouquid A, Gibelli L, Outada N (2017) A quest towards a mathematical theory of living systems. Springer, Berlin

Bennett EM, Cumming GS, Peterson GD (2005) A systems model approach to determining resilience surrogates for case studies. Ecosystems 8(8):945–957

Ceballos G, Ehrlich PR, Dirzo R (2017) Biological annihilation via the ongoing sixth mass extinction signaled by vertebrate population losses and declines. Proc Nat Acad Sci 114(30):E6089–E6096

Cohen JE (2004) Mathematics is biology's next microscope, only better; biology is mathematics' next physics, only better. PLoS Biol 2(12):e439

Darwin F (ed) (1887) The life and letters of Charles Darwin. Vol I. Free Public Domain Book from the Classic Literature Library Page 21. https://charles-darwin.classic-literature.co.uk/the-life-and-letters-of-charles-darwin-volume-i/ebook-page-21.asp. Accessed 18 Jan 2020

Downing AS, van Nes EH, Mooij WM, Scheffer M (2012) The resilience and resistance of an ecosystem to a collapse of diversity. PLoS One 7(9):e46135

Dunn RR, Harris NC, Colwell RK, Koh LP, Sodhi NS (2009) The sixth mass coextinction: are most endangered species parasites and mutualists? Proc R Soc B: Biol Sci 276(1670):3037–3045

Ellis JD, Evans JD, Pettis J (2010) Colony losses, managed colony population decline, and Colony Collapse Disorder in the United States. J Apic Res 49(1):134–136.

Fife PC (1979) Mathematical aspects of reacting and diffusing systems. Springer, New York

FitzHugh R (1969) Biological engineering, ch. Mathematical models of excitation and propagation in nerve. McGraw-Hill, New York, pp 1–85

Flannigan MD, Stocks BJ, Wotton BM (2000) Climate change and forest fires. Sci Total Environ 262(3):221–229

Friedman A, Hu B (2007) Uniform convergence for approximate traveling waves in linear reaction-hyperbolic systems. Arch Ration Mech Anal 186:251–274

Grimm V, Wissel C (1997) Babel, or the ecological stability discussions: an inventory and analysis of terminology and a guide for avoiding confusion. Oecologia 109(3):323–334

Hallmann CA, Sorg M, Jongejans E, Siepel H, Hofland N, Schwan H, Stenmans W, Müller A, Sumser H, Hörren T, Goulson D, de Kroon H (2017) More than 75 percent decline over 27 years in total flying insect biomass in protected areas. PLoS One 12(10):e0185809. https://doi.org/10.1371/journal.pone.0185809

Hardin, G (1960) The competitive exclusion principle. Science 131(3409):1292–1297

Hills TT, Todd PM, Lazer D, Redish AD, Couzin ID (2015) The cognitive search research group. Exploration versus exploitation in space, mind, and society. Trends Cogn Sci 19(1):46–54

Hodgkin AL, Huxley AF (1952) A quantitative description of membrane current and its application to conduction and excitation in nerve. J Physiol 117:500–544

IPBES (2019) Media release: nature's dangerous decline 'unprecedented'; species extinction rates 'accelerating'. intergovernmental platform on biodiversity and ecosystem services. Retrieved 6 May 2019

Janssen A, van Gool E, Lingeman R, Jacas J, van de Klashorst G (1997) Metapopulation dynamics of a persisting predator-prey system in the laboratory: time series analysis. Exp Appl Acarol 21(6–7):415–430

Janzen FJ (1994) Climate change and temperature-dependent sex determination in reptiles. Proc Nat Acad Sci 91(16):7487–7490

Kallimanis AS (2010) Temperature dependent sex determination and climate change. Oikos 119(1):197–200

Karlin S (1986) Evolutionary processes and theory. Academic Press, New York

Karlin S, Altschul SF (1990) Methods for assessing the statistical significance of molecular sequence features by using general scoring schemes. PNAS 87(6):2264–2268

Karsai, I, Kampis G (2010) The crossroads between biology and mathematics: the scientific method as the basics of scientific literacy. BioScience 60(8):632–638

Karsai, I, Knisley D., Knisley J., Yampolsky L and Godbole A. (2011). Mentoring interdisciplinary undergraduate students via team effort. CBE—Life Sci Educ 10:250–258

Kirkpatrick S, Gelatt CD, Vecchi MP (1983) Optimization by simulated annealing. Science 220(4598):671–680

Larrañaga P, Calvo B, Santana R, Bielza C, Galdiano J, Inza I, Lozano JA, Armañanzas R, Santafé G, Pérez A, Robles V (2006) Machine learning in bioinformatics. Briefings Bioinf 7(1):86–112

Lee KS, Geem ZW (2005) A new meta-heuristic algorithm for continuous engineering optimization: harmony search theory and practice. Comput Methods Appl Mech Eng. 194(36–38):3902–3933

Levin SA (ed) (1992) Mathematics and biology: the interface, challenges and opportunities. University of North Texas Libraries, UNT Digital Library. https://digital.library.unt.edu/ark:/67531/metadc1187962/. Accessed Jan 18 2020

Levin A, Simon R, Carpenter S (2012) The Princeton guide to ecology. Princeton University Press, Princeton

Mather WH, Hasty J, Tsumring LS, Williams RJ (2011) Factorized time-dependent distributions for certain multiclass queueing networks and an application to enzymatic processing networks. Queueing Syst 69:313–328

May RM (2004) Uses and abuses of mathematics in biology. Science 303:790–793

May R, McLean A (2007) Theoretical ecology: principles and applications, 3rd edn. Oxford University Press, Oxford, pp 98–110

McGill BJ, Dornelas M, Gotelli NJ, Magurran AE (2015) Fifteen forms of biodiversity trend in the Anthropocene. Trends Ecol Evol 30(2):104–113

Meadows DH, Meadows DL, Randers J, Behrens WW (1972). The limits to growth; a report for the club of Rome's project on the predicament of mankind. Universe Books, New York. ISBN: 978-0-87663-165-2

Niemelä, E, Makinen TS, Moen K, Hassinen E, Erkinaro J, Lansman M, Julkunen M (2000) Age, sex ratio and timing of the catch of kelts and ascending Atlantic salmon in the subarctic River Teno. J Fish Biol 56(4):974–985

Olton, DS, Schlosberg P (1978) Food-searching strategies in young rats: win-shift predominates over win-stay. J Comp Physiol Psychol 92(4):609–618. https://doi.org/10.1037/h0077492

Parks PC (1992) A. M. Lyapunov's stability theory—100 years on. IMA J Math Control Inf 9(4):275–303

Peli G, Nooteboom B (1999) Market partitioning and the geometry of the resource space. Am J Soc 104(4):1132–53

Reed MC (2015) Mathematical biology is good for mathematics. Not AMS 62(10)

Shou W, Bergstrom CT, Chakraborty AK, Skinner FK (2015) Theory, models and biology. Elife 14(4):e07158

Sturmfels B (2005) Can biology lead to new theorems? Annual report of the Clay Mathematics Institute, 13–26

Thompson D'Arcy W (1917) On growth and form. Cambridge University Press, Cambridge

Turing AM (1952) The chemical basis of morphogenesis. Philos Trans R Soc B 237:37–72

Turner GM (2008) A comparison of The Limits to Growth with 30 years of reality. Global Environ Change 18(3):397–411

Whitley D, Watson JP (2005) Complexity theory and the no free lunch theorem. In: Search methodologies. Springer, Boston, pp 317–339

Wright S (1984) Evolution and the genetics of populations: genetics and biometric foundations, vol 1–4, U. Chicago Press, Chicago

Chapter 2
The Importance of Life History and Population Interactions in Population Growth

Abstract Population interactions are the centerpiece of many ecological studies. Especially the roles of competition and predation on populations on dynamics and evolution have remained a hot topic since Lotka and Volterra's classical works. However, understanding how these population interactions are interacting with each other has remained elusive. We tested the hypothesis of Gurevitch that predation will stabilize an ecosystem, because it should decrease competition via removing resource consumers from the system. In a top-down model, we show that implementing even just a few life history parameters, the population will stay below their carrying capacity, which automatically dampen the effect of competition. Therefore, a decrease of competition can happen without predators. Our bottom-up, agent-based model predicted then that predation actually fosters the removal of one competing prey species from the system. A longer coexistence of preys can only be observed if they occupy different niches. In this case, the effect of predation is not only more moderate, but it will increase the survivability of the predators as well. We emphasize the importance of implementing life history parameters (e.g., mating success, sex ratio, density-independent death) into the population models to find more reliable predictions.

2.1 Background

Early models tended to assume that organisms live in large panmictic populations (i.e., perfectly mixed, in the sexual and genetic sense) and in a homogeneous environment with a one-to-one sex ratio (or without males) with a completely random mating, and with a Mendelian segregation of the genes that are thus assumed to be independent (Colegrave 1997; Fisher 1930; Eshel and Feldman 1982; Lessard 1990). However, in nature, populations are limited in space and the resource availability, therefore, population structure and environmental factors should have profound consequences for the ecology and evolution of organisms. Recent studies pointed out that populations possibly reside far away from any equilibrium—still

© Springer Nature Switzerland AG 2020

I. Karsai et al., *Resilience and Stability of Ecological and Social Systems*,
https://doi.org/10.1007/978-3-030-54560-4_2

they are resilient and can cope with some perturbations. The observed population size does not necessarily stay close to the carrying capacity of the environment (Rohde 2006).

Williams (1966) represented a turning point from the early widespread concept of population level fitness (Wynne-Edwards 1962), stressing the importance of individual level selection. Williams (1966) asserted two predictions: (1) Male-biased sex ratios should evolve whenever the population density needs to be reduced; (2) Female-biased sex ratios should evolve whenever the survival of the population is enhanced by an increased growth rate. He argued that there is no evidence that the sex ratio is changing in this way, because natural selection must not be acting to favor the group but the individual. Population size changes in the course of the evolution via increasing the carrying capacity either by extending the resource availability in the habitat or by using the resources more efficiently.

Historically, population theories remained strongly female-focused (perhaps because only females bring offspring) and the role of males in the population dynamics of animal species has been largely neglected (Mysterud et al. 2002). This view started to change in the last decade. For example, Engen (Engen et al. 2003) showed that fluctuations in the sex ratios both in monogamous and polygamous systems add an important component to the demographic variances. Change in the sex ratio also has some interesting indirect effects. For example, it is well documented that the presence of males induces estrus in females. Skewed sex ratios or populations with many young males thus result in a delayed or less synchronous calving of females (Mysterud et al. 2002). An explicit modeling of males and females is required in order to study the population-dynamic consequences of these aspects. The "asexual" models that still prevail in theoretical studies of population dynamics provide only a very limited understanding of realistic dynamics (Berec and Boukal 2004).

Local male competition (Hamilton 1967) is most commonly mentioned as an explanation for the occurrence of biased sex ratios. This has triggered many studies focusing on male mating success, but most models that were used to make sense of the observed population growth did rely only on density-dependent parameters. However, individuals also die in considerable extent due to density-independent reasons (e.g., age-related death). Such density-independent mortality rates were used in some classical ecological models like in predator–prey models (Lotka 1925; Volterra 1926) and also in some more recent models (e.g., Runge and Marra 2005; Berec and Boukal 2004; Berec et al. 2001). Slobodkin (1961) modified the basic Lotka–Volterra model by including a non-selective removal factor that was crucial to explain Gause's (1934) observations on competitive coexistence. Contrary to these works, most population models neglect the density-independent mortality (e.g., Ricker 1954; see also the overview in Brännström and Sumpter 2005). This neglect is especially critical, because mortality rates are also sex-related (which is another important function, neglected most of the time). It is known that male mortality rates are typically higher than those of females (Owen-Smith 1993), even when males and females have a similar body size (Gaillard et al. 1993).

It is also very rare that a population is isolated and has no interaction with other species. In many cases the intraspecific interactions dominate, because the conspecific individuals have the same needs and therefore they occupy the same ecological niche. However, in most cases every population faces a plethora of interactions with other populations and some of these interactions could be stronger than the intraspecific interactions. The strength and nature of these interactions can show a large variation in time and space. For example, if predators are scarce, they provide a very little predation pressure on the prey population, but this can change considerably when the predators are abundant or they switch their focus on the given prey species. This varying strength of the interactions can result in interesting dynamics. We want to explore the effect of these interactions.

2.1.1 Model Formulation

We present a model based on the Verhulst sigmoid growth (Verhulst 1845), but we have also added three very basic life history parameters (sexes exist—the sex ratio can be calculated—having different mortalities and mating success). This small change in approach will have a profound effect on population dynamics. For example, we will show that density-independent mortality intensifies the negative effects of sub-optimal adult sex ratios (ASR) in a way that significantly lowers the abilities of a population to survive in conditions of strong interspecific competition. We stress that considering a sex-related density-independent mortality factor is important to understand sex ratio evolution and the population dynamics of all sexual species. Interaction of basic life history parameters may play an even more crucial role in those species in which changing environmental conditions can affect sex determination much more than do other life history characteristics (Charnov and Bull 1977).

We start with the classical Verhulst equation (where N is the population size, r is the reproduction rate, and K is the carrying capacity). The original model considers neither the existence of sexes nor the density-independent death of organisms:

$$\frac{dN}{dt} = r(1 - \frac{N(t)}{K})N(t) \qquad (2.1.1.1)$$

Density-independent death is different from the density-dependent death rate that is embedded in r, because this describes those deaths, which happen without considering population densities. Including density-independent death is simply adding a negative (removal) term, which can serve also to describe the external exploitation of the population (fishing, hunting, if any) as well as age rated deaths:

$$\frac{dN}{dt} = r(1 - \frac{N(t)}{K})N(t) - \mu N(t) \qquad (2.1.1.2)$$

This term has a far reaching importance, because it keeps the population below the carrying capacity and this, in turn, also moderates competition (details on this come later). In reality, μ depends on the sex of the individuals, therefore first we have to specify that the population is not homogenous, but it consists of males (N_m) and females (N_f), thus $N = N_m + N_f$. From the birth of males and females we can calculate the birth sex ratio (BSR), but we will work with α, which denotes the adult sex ratio (ASR):

$$\alpha = \frac{N_f(t)}{N_m(t) + N_f(t)} \tag{2.1.1.3}$$

For the mating, we assume that every male fertilizes at most λ females and that every female mates with just one male. This is a strong simplification of the pair formation problem studied in detail by Berec and Boukal (2004). The tenet has the consequence that the sex ratio and the male success affect how many females are able to bear offspring, but it still retains the oversimplification of random mating and the assumption of an equal chances for success. Here the goal was not to make an extensive study of the different aspects of male success (mate search, mate choice, divorce behavior, etc.), but to examine the compound effects of these parameters. Based on our simple assumptions, we calculate the number of successfully mating females ($F_m(t)$) as follows:

$$F_m(t) = \min \left(\begin{matrix} N_f(t) \\ \lambda N_m(t) \end{matrix} \right) \tag{2.1.1.4}$$

This means that the number of fertilized females has an upper bound of either $N_f(t)$ or $\lambda N_m(t)$. While the "min" function is not an elegant mathematical description to use in aggregated models, this is actually a quite realistic interpretation biologically.

Now introducing the sexes and male success, we can rewrite the modified Verhulst model for population growth as this:

$$\frac{dN_f}{dt} = r\alpha(1 - \frac{N(t)}{K})F_m(t) \tag{2.1.1.5}$$

$$\frac{dN_m}{dt} = r(1 - \alpha)(1 - \frac{N(t)}{K})F_m(t) \tag{2.1.1.6}$$

At this point we have separate equations for the males and females and the reproduction will come ultimately from the fertilized females ($F_m(t)$). Term r contains density-dependent death by definition. However, in real populations it is very common that death is more or less independent from how closely a given population has approached the carrying capacity K. As mentioned before, introducing a density-independent mortality (μ) reflects the summation of random (accidental) death events, harvesting or age-related death. Using different values of μ for the two sexes allows us to model sex-specific mortality rates. Our final equations for males and females thus are

$$\frac{dN_f}{dt} = r\alpha(1 - \frac{N(t)}{K})F_m(t) - \mu_f N_f(t) \qquad (2.1.1.7)$$

$$\frac{dN_m}{dt} = r(1 - \alpha)(1 - \frac{N(t)}{K})F_m(t) - \mu_m N_m(t) \qquad (2.1.1.8)$$

These equations also assume that reproduction and mortality operate concurrently, that is, newborns do not die at the same time step when they are born and the adults that will die have a full opportunity to reproduce at that time step. Due to a strong negative feedback in these equations, we will see equilibria emerge for both populations, which we can calculate by setting the rate of change (left hand side) to zero:

$$N^*_{total,f} = K(1 - \frac{\mu_f N_f(t)}{r\alpha F_m(t)}) \qquad (2.1.1.9)$$

$$N^*_{total,m} = K(1 - \frac{\mu_m N_m(t)}{r(1 - \alpha)F_m(t)}) \qquad (2.1.1.10)$$

Our equations show clearly that the predicted total equilibria are both below K as long as μ_f and μ_m are above zero. Thus, the model predicts that density-independent mortality of either sexes keeps the population size below the carrying capacity of the environment. This means that a fraction of resources, that is, available in the habitat, is never exploited by the modeled population due to species-intrinsic limitations of growth. The assumption of simultaneous birth and density-independent death keeps the population below carrying capacity. This effect of an intrinsic population limitation by density-independent death and by sexual reproduction is of high significance in ecological considerations (Schmickl and Karsai 2010). How far populations are intrinsically limited below K depends on the values of α, $F_m(t)$, and thus also on λ.

By equating the equilibria equations, we can search for the ratio of males to females at the point in time when both sexes have reached their equilibrium population size. This yields

$$\frac{N_f(t)}{N_m(t)} = \frac{\mu_m \alpha}{\mu_f(1 - \alpha)} \qquad (2.1.1.11)$$

The expression shows that the ratio of females to males increases with a decrease of μ_f and with the increasing values of α. However, the final adult sex ratio (ASR) is independent from λ and r, and thus also independent from $F_m(t)$. Deeper analytical derivations would be difficult, especially due to the nonlinear and non-continuous function $F_m(t)$, therefore we will next turn to the use of numerical methods. This analysis will show us how the combination of parameters μ_f, μ_m, λ, and α affects the emergence of un-exploited resources in the habitat.

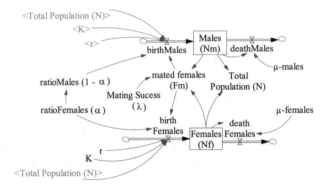

Fig. 2.1 Stock and flow representation of the model. The two boxes represent the population size of males and females, respectively. Inflows are the density-dependent reproductions and outflows are the density-independent deaths (double arrows). Other components (plain text with single arrows) indicate auxiliary system variables and constants as well as their dependencies. Basic parameters (if not mentioned otherwise): $\mu_f = \mu_m, = 0.1$, $\lambda = 1$, $K = 10,000$, $N_f(t = 0) = N_m(t = 0)) = 1$, $r = 2$, $\alpha = 0.5$. Reprinted from Schmickl and Karsai (2010) with permission from Elsevier

2.1.2 Simulation of Population Growth and Main Predictions

To investigate the effect of the three life history parameters on population growth, we implemented the model in Vensim (Fig. 2.1). The model predicts that although varying the sex ratios ($\mu_f = \mu_m, = 0.1$, $\lambda = 1$) will affect the growth speed and the number of males and females, but not the total population size, which will reach the total carrying capacity (Fig. 2.2).

When density-independent mortality was included, however, it immediately generated interesting patterns and the population growth became much more complex. In general, the total size of the populations never reached the level of carrying capacity and an overshoot in the dynamics emerged for the sex that had a larger mortality rate. Adjusting the sex ratio was able to compensate for the sex dependent mortality, which is important if one male can fertilize only one female ($\lambda = 1$) (Fig. 2.2).

As mentioned above, the success of matings is rarely studied and most models assume that the presence of a few males ensures that all females get pregnant and every female will deliver the maximum number of offspring. We know that this general assumption is rather an exception, however, and it is not valid when parental care involved, animals are territorial, animals have mass lekking courtships[1] or the species is scarce (Allee effect). The effect of male success on final population size is extensive, especially at biased sex ratios (Fig. 2.3). Larger male success ensures larger population sizes even in case of biased sex ratios, but small

---

[1] https://en.wikipedia.org/wiki/Lek_mating.

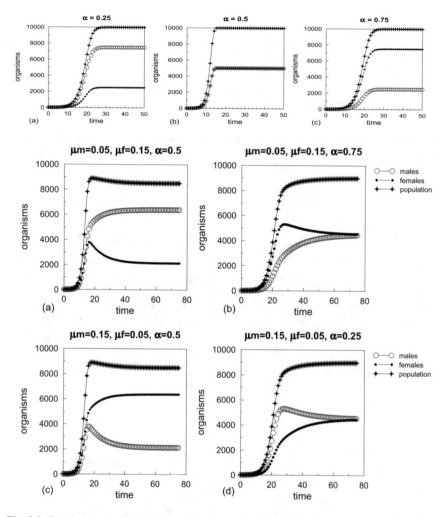

Fig. 2.2 Population growth as the function of sex ratio and density-independent mortality. Male success is constant ($\lambda = 1$: one male fertilizes only one female). Upper row: density-independent mortality switched off ($\mu_f = \mu_m, = 0$) and only the sex ratio was varied. Middle and bottom rows: density-independent mortality and sex ratio were both varied. Reprinted from Schmickl and Karsai (2010) with permission from Elsevier

male success even in male-biased populations provides smaller total population sizes altogether.

Combining the effect of all parameters ($\mu_f = \mu_m, \lambda, \alpha$), the resulting population growth has a very different characteristics from the well-known Verhulst-type sigmoid growth. One of the main findings was that the populations can only rarely reach the carrying capacity (Fig. 2.4). This has an important consequence on population interactions, which we will explore in the next chapter. Here we want

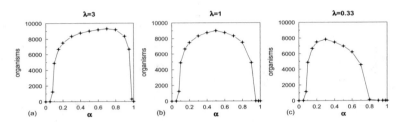

Fig. 2.3 The effect of male success on the total population size as a function of sex ratio. Density-independent mortality was kept constant ($\mu_f = \mu_m, = 0.1$). Reprinted from Schmickl and Karsai (2010) with permission from Elsevier

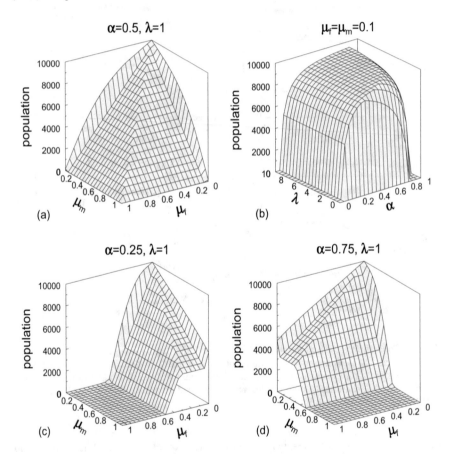

Fig. 2.4 The effect of sex ratio, male success, and density-independent mortality on population size at $t = 1000$ steps. (**a**) Unbiased sex ratio ($\alpha = 0.5$), standard male success ($\lambda = 1$) with variable density-independent mortalities were sampled. (**b**) Fixed density-independent mortalities, α and λ were sampled. (**c**) Male-biased sex ratio ($\alpha = 0.25$), standard male success ($\lambda = 1$), density-independent mortalities were sampled. (**d**) Female-biased sex ratio ($\alpha = 0.75$), standard male success ($\lambda = 1$), density-independent mortalities were sampled. Reprinted from Schmickl and Karsai (2010) with permission from Elsevier

to emphasize that the combination of these important life history parameters has a profound effect on population growth patterns and that the predictions are different from the classical and well-known work of Verhulst (1845). Now the populations can die out and in general have a much lower population size than can be expected from the carrying capacity K (Fig. 2.4).

So, while this model is still an abstract one, it had already several interesting predictions about real systems. For example, the Allee effect that describes the population extinction at low densities due to difficulties in mating (Berec and Boukal 2004; Engen et al. 2003; Allee et al. 1949), emerged automatically in this model when strongly biased sex ratios cause a lack of mating possibilities and this, in turn, leads to extinction or trifling population sizes. While the Allee effect has been known for some time (Dennis 2002), we still lack a quantitative understanding of the mechanisms affecting its population dynamics (Fowler and Baker 1991).

Density-independent mortality (often neglected in population models) strongly enhances the negative influence of unbiased sex ratios and inefficient pairing. We can assume that natural selection minimizes the intrinsic limitations of growth by decreasing density-independent mortality (μ_f, μ_m), by increasing male success (λ), and by adjusting the sex ratio (α), so that populations grow as closely as possible to the carrying capacity. However, density-independent mortality is largely depending on factors that are difficult to adapt (e.g., age, fatal accidents).

In this simple model we did not include any further details on population age, structure, or demographic stochastic factors. There is no doubt, however, that adding more life history parameters would result in even more realistic and more detailed models. Incorporation of these factors and similar processes may shed further light on ecological and evolutionary implications of population dynamics. We stress that future population models should, at least, incorporate sex ratio, male success, and density-independent mortality to make more plausible predictions of the population dynamics in gender-structured populations (Schmickl and Karsai 2010).

2.2 Interactions Between Populations

2.2.1 The Importance of Population Interactions

In real ecosystems, competitive interactions do not occur only horizontally between competing species but they also have a profound effect on trophic levels below and above that of the competitors (Connell 1971; Holt 1985; Chesson and Huntley 1997; Chesson 2000). One of the first early hypotheses, provided by (Hairston et al. 1960) and supported by many field experiments later, claimed that predators and plants are negatively affected by interspecific competition more extensively than herbivores. Gurevitch et al. (2000) concluded that in general, predation may act to reduce the intensity of competitive interactions. However, there are case studies that show a wide variety of effects of predation (increase, decrease, or no effect) on interspecific competition (Chase et al. 2002). This diversity of outcomes in real systems contrasts with the simple theoretical and also with many short-term empirical results regarding impacts on different fitness components (Chase et al. 2002). This is partly due to some confusion in the literature about both the meaning of the terms (such as "intensity") and the exact conditions required for each of these outcomes. For example, if the presence of a predator can decrease both competing prey population's sizes considerably, then resources will not be the limiting factor for the growth of these species, thus the effective competition between both prey species will decrease, and this in turn will increase the likelihood of their coexistence (Connell 1971). However, even this view is somewhat oversimplified: the diversity of potential effects of predators arises because coexistence depends on other factors as well, such as the ratio of interspecific effects to intraspecific effects and on how this ratio depends on still further facts such as resource availability or the exact mechanisms of predation (Chase et al. 2002).

The analysis of (Chase et al. 2002) of ecological interactions is based upon top-down (stepwise design) models. The authors stress that the major simplifications of the top-down models (such as homogenous populations, lack of time lags, constant life history parameters) could be violated in most natural systems. They also emphasize that new theories should not concentrate solely on adding more details into older models to remedy these issues, but rather find new ways of implementing new models. We agree that some of the limitations and assumptions of the top-down models may instigate that model predictions are different from field data. Top-down phenomenological models usually do not provide a clear mechanistic basis of the processes at the individual level. Population dynamics, in principle, result from local interactions and behaviors of a large number of individuals.

Alternative routes to the modeling of population interactions include interaction networks (Rabajante and Talaue 2015), cellular automata (Silvertown et al. 1992), as well as individual-based (agent-based) modeling (DeAngelis et al. 1992; DeAngelis and Grimm 2014). Bottom-up (disaggregated design) models, such as individual- or agent-based models, are following a model design based on a set of proximate mechanisms. In contrast, macroscopic top-down models such as the original Lotka–Volterra models (Eqs. 2.2.2.12, 2.2.2.13) are built around an observed macroscopic system behavior, which abstracts away all the microscopic processes by subsuming them into singular macroscopic system parameters such as carrying capacities and competitive coefficients. These parameters are, however, emergent properties of the actual system (and hence in reality not constants) and implicitly arise at runtime in the bottom-up models (Schmickl and Crailsheim 2006; Karsai and Kampis 2011; Grimm and Railsback 2005; Bousquet and Le Page 2004; DeAngelis and Mooij 2005). Also, in bottom-up models, spatial distributions as well as the spatiotemporal behaviors (e.g., locomotion) of the organisms can be explicitly modeled, whereas by contrast, the top-down models usually assume completely homogeneous distributions (i.e., ideally mixed populations) in the mean-field approach they are built upon.

The famous Lotka–Volterra equations have been extensively studied and extended for more biological realism to include, for example, stage structure

(Xu et al. 2005), the effect of predators (Chase et al. 2002), sex ratio, density-independent mortality (Schmickl and Karsai 2010), and more (for a review see Thieme 2003; Ahmad and Stamova 2013). In this chapter, we present a simple example on how the competition model changes when we implement some of the life history parameters described above (see also Schmickl and Karsai 2010) and then we will focus on how competition and predation interactions in fact "interact" in a simple ecosystem.

Here we present both top-down and bottom-up approaches to the modeling of population interactions. These are minimalistic and truly abstract models of a very simple ecosystem. We call attention to the inherent and important properties of natural (and bottom-up) systems, namely the role of population fluctuations, which will emerge automatically in such a system and add a new dimension to the interactions. Without having to implement carrying capacities and competition coefficients, we show how competition and predation can occur and also that agent-based models can have reasonable predictions.

2.2.2 A Top-down Population Interaction Model

The classical works of Lotka (1925) and Volterra (1926) have focused on competitive (−−) and predatory (+−) interactions. We chose to discuss these interactions in the present chapter, because of their importance and universality. Interestingly, Lotka and Volterra described the two interactions with different population growth models. Their competition model is an extension of the earlier Verhulst model (1845), where intra- and interspecific competitions are combined into a pair of coupled equations:

$$\frac{dN_1}{dt} = r_1 N_1 [1 - \frac{N_1 + \alpha_{21} N_2}{K_1}]$$

$$\frac{dN_2}{dt} = r_2 N_2 [1 - \frac{N_2 + \alpha_{12} N_1}{K_2}] \qquad (2.2.2.12)$$

Here N is the population size, t is time, K is the carrying capacity, r is the intrinsic rate of increase, and α is the competition coefficient for population 1 and 2, respectively.

The prey–predator system, which describes the case when members of one population consume the individuals of another population, is based on an exponential population growth model combined with mass action law:

$$\frac{dN_1}{dt} = r_1 N_1 - c N_1 N_2$$

$$\frac{dN_2}{dt} = -r_2 N_2 + g c N_1 N_2 \qquad (2.2.2.13)$$

Here N_1 is the prey population, which grows exponentially, but this growth is decreased by a mass action term, which contains the density of both populations and a proportionality constant c, which in turn specifies the fraction of encounters that results in prey consumption. The predator population N_2 has a decay term describing that their population will decline if the prey is not present, but if there is prey, a fraction (g) of the consumed prey (cN_1N_2) is converted into new predators.

Above we have emphasized that some basic life history parameters (such as sex, density-independent death, male success) should be included in the growth models of single populations, because even with this minimal level of added extra realism, these parameters will bring new insights into understanding population growth. One of our main finding was that populations may not reach the carrying capacity of the environment due to density-independent death. In that case, free resources are always present. This finding is especially important for understanding interspecific competition, which generally implies that free resources are rare or nonexistent: competition is assumed to be a struggle for the limited resources. To study the implications of our population growth model, we extended this model into a competition model (that can be downloaded[2]).

In the example below, two species compete and both species have males and females, therefore we have four equations to describe the competitive interactions:

$$\frac{dN_{f,1}}{dt} = r_1\alpha_1(1 - \frac{N_1(t)}{K_1} - \frac{b_{12}N_2(t)}{K_1})F_{m,1}(t) - \mu_{f,1}N_{f,1}(t) \qquad (2.2.2.14)$$

$$\frac{dN_{m,1}}{dt} = r_1(1 - \alpha_1)(1 - \frac{N_1(t)}{K_1} - \frac{b_{12}N_2(t)}{K_1})F_{m,1}(t) - \mu_{m,1}N_{m,1}(t) \qquad (2.2.2.15)$$

$$\frac{dN_{f,2}}{dt} = r_2\alpha_2(1 - \frac{N_2(t)}{K_2} - \frac{b_{21}N_1(t)}{K_2})F_{m,2}(t) - \mu_{f,2}N_{f,2}(t) \qquad (2.2.2.16)$$

$$\frac{dN_{m,2}}{dt} = r_2(1 - \alpha_2)(1 - \frac{N_2(t)}{K_2} - \frac{b_{21}N_1(t)}{K_2})F_{m,2}(t) - \mu_{m,2}N_{m,2}(t) \qquad (2.2.2.17)$$

Here $N_{f,i}$, $N_{m,i}$ describe the population sizes of species i for females and males, respectively. Parameters are the same as in the population growth model (2.1.1.7, 2.1.1.8): r: reproductive rate; K: carrying capacity; μ: density-independent death rate; α: sex ratio; λ: male success; F_m: number of fertilized females, described as

$$F_{m,i}(t) = \min\begin{pmatrix} N_{f,i}(t) \\ \lambda N_{m,i}(t) \end{pmatrix} \qquad (2.2.2.18)$$

The linkage between the two populations (i.e., the strength of interspecific competition) depends on the two parameters b_{12} and b_{21}. For simplicity we assume

---

[2]https://sites.google.com/site/springerbook2020/chapter-2.

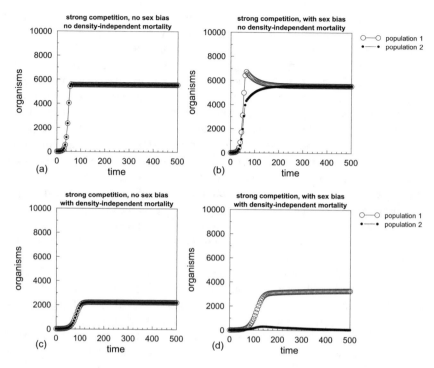

Fig. 2.5 Predicted population dynamics of two competing species with different values for sex ratio and density-independent mortality. For all simulations the same parametrization $N_{f,i}(0) = 1$, $N_{m,i}(0) = 1$, $\lambda_i = 1$, $r_i = 0.5$, $K_i = 10{,}000$, $b_{12} = b_{21} = 0.8$ was used. Other parameters have been changed between the panels: (**a**) $\alpha_1 = 0.5$, $\alpha_2 = 0.5$, $\mu_{m,1} = \mu_{f,1} = \mu_{m,2} = \mu_{f,2} = 0$; (**b**) $\alpha_1 = 0.55$, $\alpha_2 = 0.6$, $\mu_{m,1} = \mu_{f,1} = \mu_{m,2} = \mu_{f,2} = 0$; (**c**) $\alpha_1 = 0.5$, $\alpha_2 = 0.5$, $\mu_{m,1} = \mu_{f,1} = \mu_{m,2} = \mu_{f,2} = 0.15$; (**d**) $\alpha_1 = 0.55$, $\alpha_2 = 0.6$, $\mu_{m,1} = \mu_{f,1} = \mu_{m,2} = \mu_{f,2} = 0.15$; With $\mu > 0$ and $\alpha \neq 0.5$ the population of the species with the less adapted sex ratio becomes extinct (case **d**). Models that discount the existence of a biased sex ratio and/or the density-independent mortality predict a coexistence at identical population levels (**a–c**). Reprinted from Schmickl and Karsai (2010) with permission from Elsevier

that the competitive ability, i.e., the use of resources, does not depend on the sex of the individuals. This is not the case for several realistic populations, but this model can also handle the sex-related differences very well. Here we only present a few examples of what the system predicts (Fig. 2.5).

In these simple runs we have assumed that $b_{12} = b_{21} = 0.8$. This corresponds to a situation where the two populations have no perfect niche overlap. Each of the two species has an access to 20% of resources that are not available for the opponent species. Without sex bias and density-independent mortality (Fig. 2.5a), both populations were predicted to coexist at a stable equilibrium, similar to what the simple Lotka–Volterra model would have predicted. By introducing density-independent mortality (Fig. 2.5c), the species are still able to coexist, but do so at a much lower level of equilibrium compared to the previous scenario. The dynamics of both sexes is symmetrical and both species are predicted to reach the same final population size. With one species having a combination of male success and sex ratio which allows a higher maximum population size (Fig. 2.5b), the final population size depends strongly on the strength of the density-independent mortality (compare a–c with b–d). With biased sex ratio, species with a sex ratio closer to 1 : 1 show a faster population growth than do the others, which leads to a short time advantage and to the emergence of an "overshoot pattern" before reaching a final equilibrium population size (see Fig. 2.5b, for the time period $50 < t < 150$). When the model also incorporates density-independent mortality, the model's predictions are fundamentally different from the previous three cases, however: One species with a sex ratio that allows for bigger population sizes and faster population growth, wins the competition and the other species, which has just a tiny little bit less adapted sex ratio, is predicted to go extinct (Fig. 2.5d).

These examples show that if we implement even just a very few new variables that describe important natural history parameters, we can get completely different predictions from those of the classical models. The classic Lotka–Volterra competition model predicted coexistence when the interspecific competition coefficients are smaller than one (describing a situation when intraspecific competition is larger than intraspecific competition). However, this prediction contradicts the famous Gause hypothesis, which claim that competitors cannot coexist in a long term. We believe this contradiction partly comes from the fact that the classical models neglected elementary life history. It is hard to imagine that all aspects of life history would be identical for two different species. And we showed above that implementing a few fundamental parameters and having small differences between the two species will result in a prediction that is more in concert with the famous Gause hypothesis, which is built upon well-controlled experiments (Gause 1934).

In the next section, instead of elaborating further the top-down models, we change the approach and provide an example for the bottom-up modeling of the same biological problem (see the Appendix for different modeling techniques). Our goal is to stay abstract and extremely simple, while refusing to hard-wire global parameters, such as a carrying capacity and competition coefficients, which play a key role in the top-down model. We believe these parameters do not really exist in nature per se, but are emergent and dynamical properties. This way we think that the organism cannot have an access to this information directly and so it will not consider the respective variables for breeding (and other) decisions. We also will focus on how competition and predation, the two most important interactions, are connected and interact in a simple ecosystem.

2.2.3   A Bottom-up Population Interaction Model

Describing a top-down model is relatively easy, because the equations give the behavior of the system directly and we only need to specify the values of the parameters. After this, the model is solvable (even if often by computation only) and is fully reproducible. Bottom-up models are much more difficult to describe and reproducing them by other scientists can be tricky. The agents have a rule set, but how this rule set is implemented and in which order, how randomization is implemented, and how information (such as the number of individuals) is collected for processing the results are all factors affecting the outcome. There exists therefore a protocol, called ODD and developed by Grimm et al. (2010), which can be used effectively to describe agent-based models to achieve good communicability and reproducibility. We used this protocol in our technical papers, but the length and the details of this protocol are not feasible for this book. We will provide a concise description for the agent-based models we use in this book and we refer to the technical papers for details. We also provide these models with a source code for the readers for experimenting.[3]

The purpose of model below is to understand the details of a simple competitive system where carrying capacity, competition, extinction, and survivability are all emergent properties of the system and not driven by top-down parameters. The model was developed using the Starlogo TNG 1.2 environment.[4] Starlogo TNG is an agent-based modeling environment with *drag-and-drop* boxes, hence it is very accessible for first time programmers and for educational purposes. The model is fully described in (Karsai et al. 2016) and here we provide the description of the core model only. The model is minimalistic in its assumptions and in the number of agent types. We want to understand the dynamics of a system containing a single producer, two different consumers, and a predator. We want to explore how the interactions of these three levels affect coexistence and the population dynamics of the given species. While the model is close to the simplicity of the "toy models," it adheres to energy conservation and, in fact, an energy flow that drives this system. Our goal is to address the lack of understanding on the consequences of interactions between predation and competition at different trophic levels.

The model simulates the life of a pond in a 3D environment with hard borders. Producer entities, the algae, capture energy from sunlight (which is the single input for the system). Algae are consumed by two algae eater species. Strictly speaking, consumption itself is a predatory interaction, and these 2 algae eater species compete for the same food. Finally, we have a predator (fish), which consumes these algae eaters. The environment is closed and homogenous (in some cases we define a gradient) and we assume that several agents can occupy a single spatial position (patches). All individuals adhere to some very simple rules. They can move, and this movement is limited per time step and it is random, which can be translated as a floating or a random walk. The individuals have no goals in their movement, in particular they are not hunting actively, but they consume energy by bumping into food (when it occupies the same spatial position). Energy use and collection are also very simple. All agents use energy per time step for their movement and staying alive and they also can convert the energy of food into their own energy. This means

---

Fig. 2.6 Starting setup of the pond model. A defined number of algae (green), 2 species of algae eaters (red and yellow) and predators of the algae eaters (black) are randomly scattered in a 3D pond with random starting energy levels. In each time step all individuals carry out their behavior set in a random order. A constant sunlight (energy) is the only input into the system

a simple energy addition for the predator and a death for the consumed. Birth and death are thus relating to the energy level of the individuals. If the energy of an individual drops to zero, then the individual dies. If the energy level reaches a high-level threshold, then the individual will reproduce by becoming two individuals, each with half of the energy of the parent (Fig. 2.6).

First we have taken only two trophic levels (algae and algae eaters) to examine the competitive interactions between the algae eaters in a dynamic environment (fish was switched off in these runs). The individual runs show a very interesting dynamics which is very different from what the continuous models predict. Namely, the system shows strong fluctuations, especially in a low input environment (Fig. 2.7, solid lines). In a low input environment (when a single algae can only absorb a small amount of energy per time step) these strong fluctuations will lead to the extinction of one or both of the algae eaters. When the environment is much more productive, fluctuations will be more moderate and long time coexistence is more probable (Fig. 2.7, dotted lines). The population fluctuations are less strong than it could be observed in the top-down Lotka–Volterra model. The reason is the existence of "space." It is naturally implemented in this model, but is lacking in the

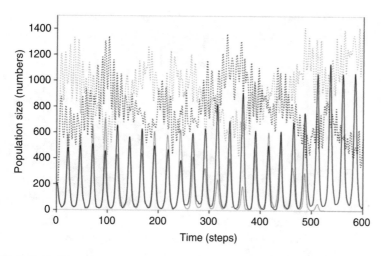

Fig. 2.7 Individual runs of the pond model. The two populations of algae eaters (black and grey) fluctuate in time. This fluctuation is stronger in case of a low energy influx (when a single algae can absorb only 0.25 energy unit per time step) and the extinction of one or both algae eaters are more probable (solid lines). In case of high energy influx (when a single algae can absorb 2.0 energy units per time step), the fluctuations are more moderate and coexistence is more probable in the long run (dotted lines). Reprinted from Karsai et al. (2016) with permission from Taylor and Francis

top-down models. The effects of space are profound. Some algae in the individual-based model are not accessible for consumption, because they are not in the vicinity of the algae eaters. In this model therefore, the algae eaters can starve and die even if there are many algae still in the pond, only farther away. These algae are in temporary refuge and they can breed, so the number of producers would not go to such a low value as we can observe in a top-down model. It is also quite rare that the algae are getting into extreme high numbers either, because the algae eaters will have an easy access to them and breed quickly when there is plenty of food. This system is not driven and constrained by a carrying capacity, but rather by the accessible energy (yet a finite carrying capacity emerges). The system dynamics is driven more by the prey–predator interactions (between the algae and the algae eaters) than the competitive interaction between the different algae eaters.

From the results of a large number of simulations, it became clear that one of the two algae eater species with identical life history parameters (all parameters identical) became extinct, but it was by a 50–50% chance, which one died out in a given run. The extinction time was dependent on the energy influx of the system (Fig. 2.8). The average time for the extinction of the first algae eater species increased threefold from 0.25 to 1.25 energy input, and then it levelled off. This shows that some extinctions happen even if the energy input is very high. The agent-based model supports the Gause hypothesis (that foresees competitive exclusion) better than do the continuous models.

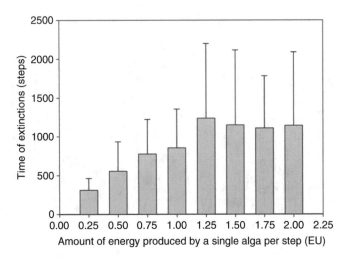

Fig. 2.8 Average times and standard deviation of the extinction of one of the consumers as a function of the energy captured by a single algae per time step. Twenty-four parallel simulations were run, each until one of the consumers has died out. To enhance competition, an extra energy loss was implemented when any two consumers collided. Reprinted from Karsai et al. (2016) with permission from Taylor and Francis

Different species generally have different life histories and in these cases the competitive interactions become even more interesting. The Lotka–Volterra competition model predicts that species with larger K and b parameters will win the competition and can drive the other species to extinction. Our individual-based model does not have these parameters, but we have the size and speed of the individuals instead. Larger size could be beneficial, because a larger body volume means larger numbers of possible interactions with food, which could result in more energy gain per time step. Larger speed means that the organism could disperse faster and has a larger probability to find pockets of food. Parameter sweeps of different combinations of the effect of size and speed revealed trade-offs (Fig. 2.9, and see Karsai et al. 2016 for more details).

When we study the effect of different life history parameters in different environments we can map the success of these combinations. We stress that this is not an evolutionary model, but a short-term ecological model, which assumes that no parameters of the species in question will change or evolve during the studied time frame. To show an example for this approach, we made algae eater 1 a bigger and slower species while algae eater 2 stayed at its standard size and speed. Again, the time needed for a species to become extinct depended upon the productivity of the environment. Similarly to what we observed when the two competing species had the same life history, the time to extinction has increased with the productivity until the algae were able to capture 1.25 unit of energy per time step. However, while the extinction time stayed high when the 2 species were identical, we see a sharp decline in the extinction time for the highest productivity environments, for

Fig. 2.9 Probability of the survival of algae eater 1 when the moving cost and size of algae eater 1 were changed while the same for algae eater 2 were set to a constant = 2. Reprinted from Karsai et al. (2016) with permission from Taylor and Francis

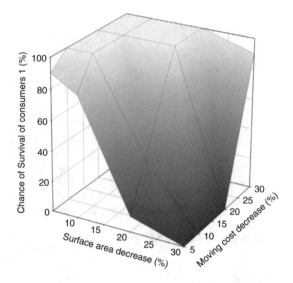

the case when the 2 species were different (Fig. 2.10). If the food is scarce (in the model, this is when the algae capture less than 1 unit energy per turn, hence they can only reproduce slowly), the algae eater populations strongly fluctuate and one of these species will hit zero (go extinct) sooner or later. In our special case, always the smaller and faster algae eater 2 will win, because it disperses faster and finds the pocket of algae with a higher probability (Fig. 2.11). However, in a high productivity environment a larger body size proves more beneficial than speed, and the slower but larger algae eater 1 will outcompete and drive species 2 to extinct.

These examples show that competition between two species can be studied without directly implementing carrying capacity and competition coefficients. Species with different life histories have a different way to cope with the environment and the competition is in fact itself an emergent property of these simple rules and coping mechanisms.

So far we have seen cases where the algae eaters were (competitive) predators of the algae. To test how coexistence and stability of this system change, we then introduced a predator (fish) that predates on the algae eaters. One of the hypotheses mentioned above comes from the Lotka–Volterra models: If the predators prey upon the competitors, then their population size decreases below their carrying capacity, and this, in turn, ensures free resources, which means that the competition between the species decreases and therefore they will coexist. In our model we introduced the predator agent (fish) in the same way as the other species. These are just "floating spheres" without a hunting behavior and consume their prey without preference or choice by bumping into any of the algae eater individuals. We also assumed that predation requires more effort than does filtering algae and we defined the quality of the predator species via its loss of energy per time step. Thus a predator species that loses only 1 energy per time step is a much more efficient predator than the species that loses, e.g., 6 energy units in a time step. This is a very abstract and simple

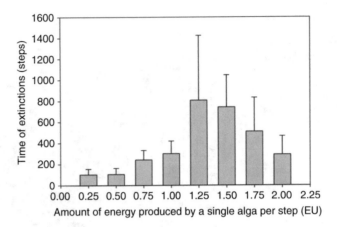

Fig. 2.10 Average time and standard deviation (24 runs) of the extinction of one of the algae eaters as a function of energy captured by a single algae per step. All parameters identical to the previous runs except that the speed (V_c) and size (S_c) of algae eater 1 ($V_{c1} = 1.5$, $S_{c1} = 3$) were different from those of algae eater 2 ($V_{c2} = 2$, $S_{c2} = 2$). Reprinted from Karsai et al. (2016) with permission from Taylor and Francis

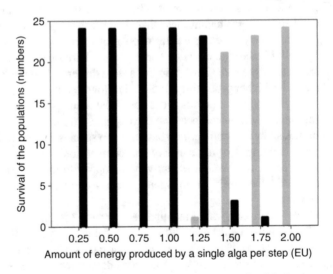

Fig. 2.11 Number of survivals (in 24 runs) of the consumer species (black: algae eater 2; grey: algae eater 1) as a function of energy produced by a single alga per step. All parameters identical to the previous runs except that the speed (V_c) and size (S_c) of algae eater 1 ($V_{c1} = 1.5$, $S_{c1} = 3$) were different from algae eater 2 again ($V_{c2} = 2$, $S_{c2} = 2$). Reprinted from Karsai et al. (2016) with permission from Taylor and Francis

solution we choose to avoid having another parameter in the model. Very efficient predators commonly wiped out the prey species and very inefficient ones died out quickly (Fig. 2.12). These results are in concert with the textbook conclusion that

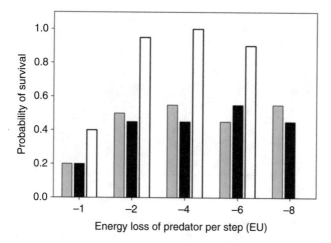

Fig. 2.12 Probability of survival of the populations as a function of predator effectiveness. Predator effectiveness was simulated as the amount of energy loss per step. Smaller values mean more effective predators. Twenty parallel simulations per parameter set were run for 1000 time steps. Grey: algae eater 1, black: algae eater 2 (with same life history), white: predators of algae eaters. Reprinted from Karsai et al. (2016) with permission from Taylor and Francis

only the mediocre predators can function in the ecosystems in the long term and the very efficient ones die out because of an over-exploitation of their prey.

Contrary to the "logical derivation form the Lotka–Volterra system" as described above, in the current model the predators foster the extinction of one of the algae eaters (Fig. 2.13). In fact, in general, their extinction is 4 times (or even more) faster than without the presence of predator (Fig. 2.8). This contradiction may come from the different dynamics of the two models. The agent-based model shows strong fluctuations, which is a characteristic also of natural habitats, but this is totally missing in the classic Lotka–Volterra competition model. In the agent-based model, the predators simply enhance the existing fluctuations, and so one of the algae eaters dies out soon. Very effective predators $(-1, -2)$ drove the extinction of one of the algae eaters significantly faster than the less effective ones (Fig. 2.13). In high efficiency predation, these habitats were unable to sustain as many predators as those with less efficient predators (Fig. 2.14). When the predators are mediocre in quality $(-4, -6)$, the habitat was able to sustain the highest number of predators and this also ensured the longest coexistence between the competitors.

Similarly, implementing a predator that preys upon the competitive consumers did not support the general conclusion by Gurevitch et al. (2000), that predation promotes the coexistence as a reducer of the intensity of competitive interactions. As we saw above, predation actually accelerates extinction of one or both competing prey species, instead of a stabilization effect. However, in nature we see a large variety of coexistent species, which are predated by many predators. This natural diversity has lots of reasons and supporting mechanisms, such as migrations, differential predation, as well as other factors and interactions, such as diseases

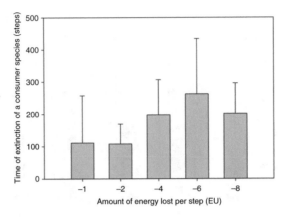

Fig. 2.13 Time it takes for one of the consumer species to become extinct (time average and standard deviation for 20 runs) as a function of predator effectiveness. The extinction times of one of the consumers were significantly different (Kruskal–Wallis H test: 43.3, $p < 0.05$, $N = 100$). Pairwise comparisons showed that this difference stemmed from the fact that -1 and −2 were significantly smaller than the other columns (while between −1 and −2 there was no significant difference). Reprinted from Karsai et al. (2016) with permission from Taylor and Francis

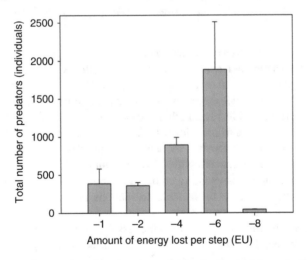

Fig. 2.14 Average of the total number of predators (initiated + born) with standard deviations as a function of predator effectiveness. Twenty parallel simulations per parameter set were run for 1000 steps. The number of predators with different efficiency was significantly different (Kruskal–Wallis H test: 79.43, $p < 0.05$, $N = 100$). Pairwise comparisons show that all columns were significantly different, except −1 and −2, and −4 and −6, respectively). Reprinted from Karsai et al. (2016) with permission from Taylor and Francis

and so on. In our abstract model system we added one simple mechanism that made our algae eaters different. We changed the movement of the algae eaters and this behavioral difference resulted in niche segregation. The implemented change

Fig. 2.15 Stratification of algae eaters as a result of preferential movement in the simulated pond. Algae (green) is most common in the middle layer where the algae eaters are less common, while algae eaters 1 (yellow) are most common in the upper and algae eaters 2 (red) in the lower layer of the water column. Predators (black) are randomly scattered. Reprinted from Karsai et al. (2016) with permission from Taylor and Francis

is in the direction of the movements: from now on, algae eater 1 prefers to move upwards and algae eater 2 downwards (Fig. 2.15). The general floating nature of the movement is still present (the individual can move to any neighboring position). First a "dice roll" tells the random amount of motion as before. Another "dice roll" has been implemented for the preference of the downward and upward movement—the number of "sides" of the "dice" determines the degree of the preference (neutral, slight, moderate, or extreme).

The effect of the niche segregation on coexistence is quite dramatic (Fig. 2.16). Even in the presence of predators, slight and moderate preference in directional movement results in a significant increase in coexistence by several times. Only an extreme degree of motion preference does not imply a significantly higher coexistence—because, due to their strong directional movement, all algae eaters were gathering at the very top and bottom of the water column. While they did not overlap and compete interspecifically with each other, the intraspecific competition was very high, since the algae eaters just use a narrow segment of the pond for feeding. This way their populations fluctuates heavily due to intermittent starvation and predation pressure from the fish, and this in turn accelerates the extinction of one or both algae eater species. A moderate directional preference not only gives the longest coexistence for the algae eaters, but it is also able to sustain the largest number of predators (Karsai et al. 2016).

In this chapter our goal was to show that while the classic Lotka–Volterra equations serve as valuable starting points, their real predictive value is low. Expanding

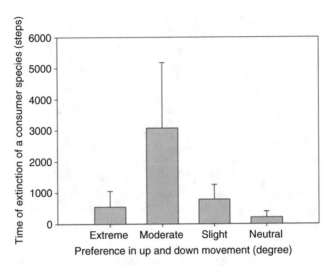

Fig. 2.16 The time until one of the consumer species goes extinct (time average and standard deviation for 20 runs) at different degrees of preference in the vertical movements. Reprinted from Karsai et al. (2016) with permission from Taylor and Francis

the equations with life history parameters will give more realistic predictions. However, making ultrarealistic models with a large number of parameters is not always the best goal for modeling, because complicated models tend to be very unstable. As Wangersky (1978) pointed out

> The more general models of theoretical biology are used to deduce the form of possible solutions, rather than to predict future states of the system being modelled.

We agree that abstract and general models play crucial role understanding interactions of basic processes, such as interaction between predation and competition, especially when field data is contradictory or shows great variety in nature.

The agent-based model we developed is also an abstract and general model, but nevertheless we designed a system where the carrying capacities and competition coefficients were not explicitly implemented in the system. These coefficients represent envelope terms of differences in many life history parameters. The benefit of agent-based models is that these differences in life history can be implemented explicitly into the system and their effects can be studied separately. For example, instead of using an explicit competition coefficient, our agent-based model relied on several interaction mechanisms at an individual level. Competition emerged as a result of these mechanisms and not determined by a constant. Similarly, carrying capacity represents the maximum number of individuals that can be sustained in the habitat. Instead of implementing a global K value, the energy exchange by interaction drove the system: producers collected energy from their environment to build up biomass, to reproduce, and to move. The generated biomass can also be transferred to the next trophic level as consumed food, yielding energy for reproduction, growth, and motion to those predatory species. Sensing and food consumption were strictly local, therefore there was no global information available to the individuals concerning the absolute amount of food in the system. Thus, investigating the same problem with two different types of modeling approaches gave us different insights in understanding the resilience and stability of competitive and predatory interactions.

References

Ahmad S, Stamova IM (2013) Lotka–Volterra and related systems. Recent developments in population dynamics. Walter de Gruyter GMBH, Berlin and Boston

Allee WC, Emerson AE, Park O, Park T, Schmidt KP (1949) Principles of animal ecology. Saunders, Philadelphia

Berec L, Boukal DS (2004) Implications of mate search, mate choice and divorce rate for population dynamics of sexually reproducing species. Oikos 104:122–132

Berec L, Boukal DS, Berec M (2001) Linking the Allee effect, sexual reproduction, and temperature-dependent sex determination via spatial dynamics. Am Nat 157:217–230

Bousquet F, Le Page C (2004) Multi-agent simulations and ecosystem management: a review. Ecol Model 176:313–332

Brännström A, Sumpter DJT (2005) The role of competition and clustering in population dynamics. Proc R Soc Lond B 272:2065–2072

Charnov EL, Bull J (1977) When is sex environmentally determined? Nature (London) 266:828–830

Chase JM, Abrams PA, Grover JP, Diehl S, Chesson P, Holt RD, Richards SA, Nisbet RM, Case TJ (2002) The interaction between predation and competition: a review and synthesis. Ecol Lett 5:302–315

Chesson P (2000) Mechanisms of maintenance of species diversity. Ann Rev Ecol Syst 31:343–366

Chesson P, Huntley N (1997) The roles of harsh and fluctuating conditions in the dynamics of ecological communities. Am Nat150:519–553

Colegrave N (1997) Can a patchy population structure affect the evolution of competitive strategies? Evolution 51:483–492

Connell JH (1971) On the role of natural enemies in preventing competitive exclusion in some marine animals and in rain forest trees. In: den Boer PJ, Gradwell GR (eds) Dynamics of populations. Center for Agricultural Publication and Documentation, Wageningen, pp 298–312

DeAngelis DL, Grimm V (2014) Individual-based models in ecology after four decades. F1000prime Rep 6:39

DeAngelis DL, Gross LJ, Boston MA (1992) Individual-based models and approaches in ecology. Chapman and Hall, New York

DeAngelis DL, Mooij WM (2005) Individual-based modelling of ecological and evolutionary processes. Ann Rev Ecol Syst 36:147–168

Dennis B (2002) Allee effects in stochastic populations. Oikos 96:389–401

Engen S, Lande R, Sæther BE (2003) Demographic stochasticity and Allee effects in populations with two sexes. Ecology 84:2378–2386

Eshel I, Feldman MW (1982) On evolutionary genetic stability of the sex ratio. Theor Popul Biol 21:430–439

Fisher RA (1930) The genetical theory of natural selection. Oxford University Press, Oxford

Fowler CW, Baker JD (1991) A review of animal population dynamics at extremely reduced population levels. Rep Int Whaling Commission 41:545–554

Gaillard J-M, Delorme D, Boutin J-M, Van Laere G, Boisaubert B, Pradel R (1993) Roe deer survival patterns: a comparative analysis of contrasting populations. J Anim Ecol 62:778–791

Gause GF (1934) The struggle for existence. Williams and Wilkins, Baltimore

Grimm V, Berger U, DeAngelis DL, Polhilld JG, Giskee J, Railsback SF (2010) The ODD protocol: a review and first update. Ecol Model 221:2760–2768

Grimm V, Railsback SF (2005) Individual-based modeling and ecology. Princeton University Press, Princeton

Gurevitch JJ, Morrison A, Hedges LV (2000) The interaction between competition and predation: a meta-analysis of field experiments. Am Nat 155:435–453

Hairston NG, Smith FE, Slobodkin LB (1960) Community structure, population control, and competition. Am Nat 94:421–425

Hamilton WD (1967) Extraordinary sex ratios. Science 156:477–488

Holt RD (1985) Density-independent mortality, non-linear competitive interactions, and species coexistence. J Theor Biol 116:479–493

Karsai I, Kampis G (2011) Connected fragmented habitats facilitate stable coexistence dynamics. Ecol Model 222:447–455

Karsai I, Montano E, Schmickl T (2016) Bottom-up ecology: an agent-based model on the interactions between competition and predation. Lett Biomath 3:161–180. https://doi.org/10.1080/23737867.2016.1217756

Lessard S (1990) Évolution du rapport numérique des sexes et modèles dynamiques connexes. In: Lessard S (ed) Mathematical and statistical developments of evolutionary theory, NATO ASI series C: mathematical and physical sciences, vol 299. Kluwer Academic Publishers, Dordrecht, pp 269–325

Lotka AJ (1925) Elements of physical biology. Williams and Wilkins, Baltimore

Mysterud A, Coulson T, Stenseth NC (2002) The role of males in the dynamics of ungulate populations. J Anim Ecol 71:907–915

Owen-Smith N (1993) Comparative mortality rates of male and female kudus: the costs of sexual size dimorphism. J Anim Ecol 62:428–440

Rabajante JF, Talaue CO (2015) Equilibrium switching and mathematical properties of nonlinear interaction networks with concurrent antagonism and self-stimulation. Chaos, Solitons Fractals 73:166–182. https://doi.org/10.1016/j.chaos.2015.01.018

Ricker WE (1954) Stock and recruitment. J Fish Res Board Canada 11:559–623

Rohde K (2006) Nonequilibrium ecology. Cambridge University Press, Cambridge

Runge MC, Marra PP (2005) Modeling seasonal interactions in the population dynamics of migratory birds. In: Greenberg R, Marra PP (eds) Birds of two worlds: the ecology and evolution of migration. Johns Hopkins University Press, Baltimore, pp 375–389

Schmickl T, Crailsheim K (2006) Bubbleworld.Evo: artificial evolution of behavioral decisions in a simulated predator-prey ecosystem. In: Nolfi S, Baldassarre G, Calabretta R, Hallam JCT, Marocco D, Meyer J-A, Miglino O, Parisi D (eds) From animals to animats, vol 9. Springer, Berlin, pp 594–605

Schmickl T, Karsai I (2010) The interplay of sex ratio, male success and density-independent mortality affects population dynamics. Ecol Model 221:1089–1097

Silvertown J, Holtier S, Johnson J, Dale P (1992) Cellular automaton models of interspecific competition for space. The effect of pattern on process. J Ecol 80:527–533

Slobodkin LB (1961) Growth and regulation of animal populations. Holt, Reinhart and Winston, New York

Thieme HR (2003) Mathematics in population biology. Princeton University Press, Princeton and Oxford

Verhulst PF (1845) Recherches mathématiques sur la loi d'accroissement de la population. Nouveaux Memoires de l'Academie Royale des Sciences et Belles- Lettres de Bruxelles 18:1–41

Volterra V (1926) Fluctuations in the abundance of a species considered mathematically. Nature 118:558–560

Wangersky PJ (1978) Lotka–Volterra population models. Ann Rev Ecol Syst 9:189–218

Williams GC (1966) Adaptation and natural selection. Princeton University Press, Princeton
Wynne-Edwards VC (1962) Animal dispersion in relation to social behaviour. Oliver and Boyd, Edinburgh and London
Xu R, Chaplain MAJ, Davidson FA (2005) Modelling and analysis of a competitive model with stage structure. Math Comput Model Dyn Syst 41:159–175. https://doi.org/10.1016/j.mcm.2004.08.003

Chapter 3
Habitat Fragmentation

Abstract Habitat destruction and fragmentation are common processes across multiple temporal and spatial scales in ecosystems. Beyond the naturally occurring perturbations (floods, fires), human society has reshaped most of the natural biomes, but the consequences of these processes are not fully understood. To study the effect of fragmentation and compartmentalization on ecosystem stability we separated habitat destruction from the fragmentation and this resulted in a better understanding of the effect of stabilization processes of a simple prey–predator system. Compartmentalization has a negative effect, especially for the predators, and they commonly became extinct even if the habitat destruction is insignificantly small. Stronger compartmentalization accelerates these negative effects. However, if the compartmentalization is paired with low traffic corridors between the compartments, the stability of the whole system could increase. If the compartmentalization and their connections do not go together with habitat destruction, then the ecosystem stability can be saved or increased even if compartmentalized. This finding gives a solid foundation for planning protected areas, especially for ensuring the maintenance of vulnerable key predators and shield prey populations from over-exploitation by the same predators we intend to save.

3.1 Background

Top-down models such as the classical Lotka–Volterra equations commonly assume that the environment is homogenous and the populations do not interact with the environment. However, this assumption never holds in nature. Populations not only interact with each other, but also interact with the environment and in fact they are part of a dynamic environment. The environment itself has a structure and this structure definitely affects the populations. Further, compartmentalization is a general biological principle of biological systems—compartmental models can address the issue in top-down modeling. Knisley et al. (2011) show an example of a compartment model of smolt migration. This model describes the 1-month-long voyage of the juvenile salmon (smolt) from their hatching ground in the Tucannon–Snake river system to the ocean. The river has 8 dams and these dams strongly

© Springer Nature Switzerland AG 2020
I. Karsai et al., *Resilience and Stability of Ecological and Social Systems*,
https://doi.org/10.1007/978-3-030-54560-4_3

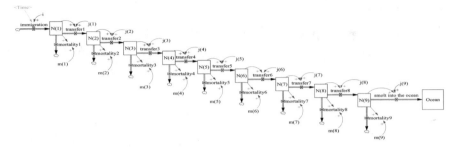

Fig. 3.1 Stock and Flow (compartment) model of the migration of the young salmon (smolt). The start of the migration happens in a given time step (Time), when i number of smolt enters in compartment 1 ($N(1)$). After this event with rates $j(i)$ the fish are moving downstream and some portion $m(i)$ will die. The double arrows describe the flows from and to the stocks (represented as squares). Single arrows describe flow of information for feedback

compartmentalize the habitat of the smolt. The river system with the hatching ground and the ocean can be modeled as 10 linked compartments (Fig. 3.1). We make the model available.[1]

In general, the change of the number of fishes in a given compartment is

$$\frac{dN}{dt} = j(k-1)N(k-1,t) - j(k)N(k,t) - m(k)N(k,t) \dots \forall k \in (2,3,\dots 9) \tag{3.1.1}$$

where N is the number of fish, k is the compartment identifier (ID), j is the transfer rate, and m is the mortality rate—except for compartment one, immigration is a single event where a number i of fish is released into $N(1)$ in a single time step. Solving the equations for each compartment, we can follow the dynamics of the fish migration over time (Fig. 3.2). The transfer rates and the mortality rates can be fitted to the nature of the river segments. For example, some river segments are larger or have "fish friendly" turbines, etc. These factors will decrease the transfer rate for that segment and also fish friendly turbines decrease the fish mortality considerably. Defining the compartments in a top-down model thus increases its predictive power and this model also can be used to predict fish yields or help planning where to install fish friendly turbines for getting more fish into the ocean. The structure of the habitat and the interaction of the population with this habitat are paramount in understanding population dynamics.

Conservationists commonly seek insights from ecological theory to select strategies of habitat management that will best maintain threatened species. Habitat fragmentation affects many ecological processes across multiple temporal and spatial scales (Schweiger et al. 2000). Despite the universal and important presence of fragmentation and the great interest expressed by conservation ecologists, there

---

[1]https://sites.google.com/site/springerbook2020/chapter-3.

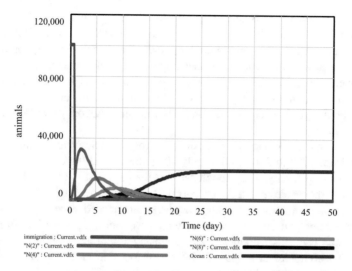

Fig. 3.2 Dynamics of the smolt populations in some of the selected compartments. One hundred thousand individuals started their migration to the ocean and propagated downstream in the river, divided into segments by dams. Using $j(k) = 0.5$ and $m(k) = 0.1$ for every compartment, 20,000 smolt migrates to the ocean in 30 days

is a difficulty in answering even some of the most important questions (Harrison and Bruna 1999). When a given habitat undergoes fragmentation, it is commonly caused by the reduction of the size and quality of the habitat. A forest habitat may become fragmented because some part of the forest is burned down for agronomy. Therefore, what was a continuous large forest before is a mosaic of patches of forest and agronomy now. Documenting the significant ecological effects of habitat loss does not convey pertinent information about the changes of the spatial structure introduced by fragmentation alone (Franklin et al. 2002). So, to assess the influence of natural habitat fragmentation, the effects of habitat loss and habitat fragmentation should be treated independently (Fahrig 2003). Investigating habitat fragmentation without habitat destruction allows us to study the effect of compartmentalization on biological systems and this could reveal the reason of several controversies of field studies. This, in turn, allows us to better evaluate the roles of ecological corridors in conservation biology.

The popularity of corridors in conservation biology stems from the intuitive relationship to their intended function by physically connecting isolated habitat fragments. Corridors are expected to increase population viability via offsetting local extinction, but also because of their restricted physical throughput the corridors regulate the extent of migration. For example, corridors can delay predator migration. When a few individuals of a heavily predated prey population migrate into a new area through a corridor, they may escape predators for a while. These areas provide a temporal refuge for the prey until the predators are also crossing through the corridor. Corridors can also be beneficial to the predators. The presence of a corridor, even if it is allowing only low levels of migration, has improved the probability of survival of a cougar population in Southern California (Beier 1993). General spatial models in ecology, including island biogeographic models (MacArthur and Wilson 1967) and metapopulation models (Levins 1969; Caswell and Cohen 1991; Hanski 1999), predict that movements between patches will increase population size and persistence. These habitat shifts are regular outcomes of how populations react to changing environments (Gyorffy and Karsai 1991; Karsai et al. 1994).

Haddad and Tewksbury (2006) reviewed major ecology and general science journals from 1997 to 2003 to find only 20 studies to test the corridors' effects on populations or diversity. They concluded that the past studies offer only a tentative support for the positive effects of corridors. They have emphasized that much more work on population and community levels are needed, especially, on studying the mechanisms and conditions under which we can expect corridors to impact populations. They also predicted an increasing importance of individual-based models that complement empirical studies by focusing on the effect of different life history parameters (Haddad and Tewksbury 2006). The study of fragmentation still seems to be controversial:

> Despite extensive empirical research and previous reviews, no clear patterns regarding the effects of habitat loss and fragmentation on predator-prey interactions have emerged (Ryall and Fahrig 2006).

Ryall and Fahrig emphasize the importance of theoretical predictions (and hence of computational models) in assessing the effects of fragmentation in predator–prey systems.

In the next section, we describe a minimalist individual-based model to study how fragmentation (without habitat loss) and the reconnection of the fragmented habitats can influence the stability of a simple predator–prey system. Individual-based modeling, as opposed to most aggregate models, allows us to use spatially explicit predation processes and simple stochastic mechanisms for the organisms to find food and new habitats. We endeavor to test the controversial ideas that exist about the role of fragmentation in a conservation context. We hypothesize that habitat fragmentation alone results in a strong detrimental effect (especially for the predator population), but connecting the fragmented habitats facilitates predator survival. We intend to demonstrate that in the presence of a high quality predator, connected fragmented habitats ensure a better chance for coexistence than does even the unfragmented original habitat. We explain why a connected fragmented habitat might thus be beneficial for the stabilization of the system, and how connections between habitat fragments are able to shield prey population from over-exploitation (Karsai and Kampis 2011).

3.2   An Individual- (Agent-) Based Model of Habitat Fragmentation

3.2.1   Description of the Model

The purpose of the model is to understand how habitat fragmentation and the re-connection of the fragments affect predator–prey systems, in particular in terms of coexistence. The original model was developed in the Netlogo 3.15 environment and it is described in full details using the ODD protocol (Grimm et al. 2006) in Karsai and Kampis (2011). The model is also deposited in the Netlogo community site.[2]

Here we describe the core of the models in non-technical terms. We built a tri-trophic model consisting of a non-mobile resource ("grass"), a prey feeding on this resource ("sheep"), and a predator ("wolf"). The dynamics of the vary-ing number of autonomous individuals is entirely controlled by the individuals' behavior. Both predators and preys are consumers, that is, they feed on biotic (replicating) resources. Prey food simply regenerates after a time, while prey itself replicates according to its individual energy budget. In our model, only consumption, reproduction, and predation are assumed at the individual organism level. Density-dependent effects and other aspects of dynamics arise as emergent consequences of the context-independent individual interactions.

The description of the model is as follows. At each time step, a given list of actions is performed in a sequential order by every individual (activated in a dynamically randomized order): move randomly in physical space, loose energy, consume available resources if possible, reproduce by chance, and die, if energy is out (Fig. 3.3) The behaviors and the results of the behaviors are converted into a single common currency called "energy." Energy is the state variable related to the well-being condition of the individual: it expresses the stored energy tokens gathered via feeding and can be used for actions. The feeding is only possible if food is in the same patch as the feeder. Upon consuming an individual, the consumer receives a certain amount of "energy," which is $Gain_{PY}$ and $Gain_{PD}$ for the prey and the predator, respectively. These parameters also relate to the effectiveness (or "quality") of the consumer. More effective consumers have a higher Gain from the same food. The consumed individual dies and is removed from the system. Reproduction of prey and predator is asexual (i.e., using the Netlogo "*hatch*" function) and occurs with a fixed probability. Upon birth, the energy reserve of the parent will be shared evenly with the offspring.

The model is spatially explicit. It consists of an area with $200x200$ spatial locations with reflecting boundaries which we assumed to be large enough to capture large-scale spatiotemporal dynamics. Each position except the borders can

---

[2]http://ccl.northwestern.edu/netlogo/models/community/HabitatFragmentation and here https://sites.google.com/site/springerbook2020/chapter-3.

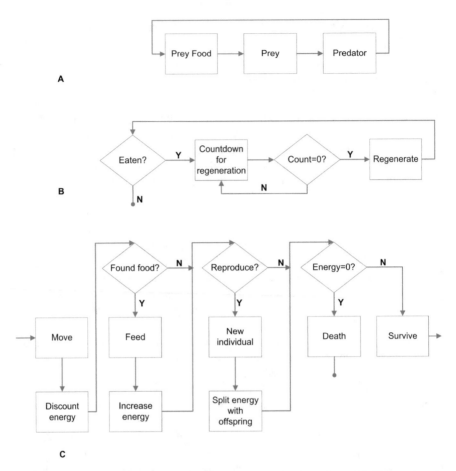

Fig. 3.3 Process overview of the model. (**a**) Basic cycle of the tri-tropic model; (**b**) growth cycle of the prey food; (**c**) activity and energy model of both prey and predator. Reprinted from Karsai and Kampis (2011) with permission from Elsevier

be empty or occupied by an arbitrary number of individuals. Prey food ("grass") is not modeled in terms of individuals, but as patches, therefore they can only exist at a certain position in a certain time or not. Grass-free patches will regenerate with a constant probability. In other words, locations are assumed to represent small finite areas where more than one individual can live, rather than spatial points of zero extension. Fragmentation is implemented by placing borders with reflecting walls into the habitat. These narrow walls, similarly to real roads, canals, fences, or other hard boundaries do not decrease the total habitat size significantly. This made it possible to study the effect of habitat fragmentation without habitat destruction.

Connections between the habitat fragments are implemented as openings in the walls, where organisms can pass through freely. These openings are large enough to permit multiple organisms to cross at the same time, but they are also

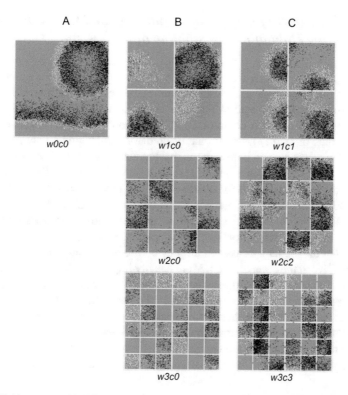

Fig. 3.4 Habitat types with different degree of compartmentalization: $wXcY$ indicates an X walls and Y connectors (openings) system (X depicts the number of divisions horizontally and vertically; Y describes the number of openings: $Y = 0$ means no opening, $Y = 1$ means each segment has one opening except those face toward the habitat border). Basic habitat setups: (**a**) unfragmented habitat (left column); (**b**) fragmented habitat (middle column); (**c**) connected fragmented habitat (right column). Light dots: prey; dark dots: predator, gray background: food for prey is present. Reprinted from Karsai and Kampis (2011) with permission from Elsevier

small enough to restrict the crossing significantly. In this study, the openings differ from real life corridors because they have practically no length or any other special properties that impede or promote migration or survival. We deliberately simplified the corridors into simple openings to fit the simplicity of the corridor to the simplicity of the movement of the animals implemented (simple random walk), so that we can focus on one single factor, namely, the different degrees of connectedness of the sub-habitats. We studied the effect of different arrangements of habitat compartmentalization by implementing different numbers of walls and openings (Fig. 3.4).

3.2.2 The Effect of Compartmentalization and Connectedness of the Habitat on the Stability of the System

Model populations behave in a qualitatively different fashion depending on the type of habitat they live in (Fig. 3.4). In the unfragmented habitat, both the prey and the predator populations tend to fluctuate heavily (Fig. 3.4a). After both prey and predator reach a peak, it is common for the prey population to collapse, especially at high values of $Gain_{PD}$ (high quality predator). Populations may nevertheless coexist for a long time and it is common that big wave-like structures sweep through the habitat (Fig. 3.4) (Kampis and Karsai 2011), and these heavy fluctuations often drive the predator population (or both predator and prey) extinct eventually.

Fragmentation compartmentalizes the habitat and, in these compartments, the respective population dynamics are played independently from each other. Commonly the predator will be extinct in many compartments quickly, but it is also possible that the predator consumes all prey before they disappear, leaving only the prey food alive in the compartment. More heavily fragmented habitat results in more of these extinctions (Fig. 3.4b). and therefore the global dynamics of the populations are different from what we have seen before. The fluctuations will disappear quickly and due to lack of predators in most compartments the surviving prey will leave at the carrying capacity of the habitat fragment (Fig. 3.5b).

When the compartments are connected, these local extinctions will still happen, but the compartments will be recolonized afterwards via the corridors (Fig. 3.4c).

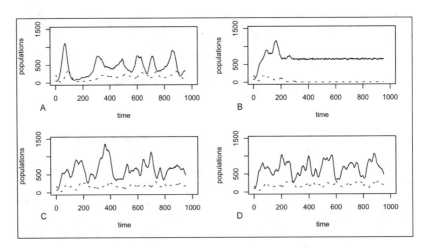

Fig. 3.5 An example of the dynamics of prey and predator populations at $Gain_{PD} = 50$. (**a**) Unfragmented ($w0c0$), (**b**) fragmented ($w2c0$), as well as two different fragmented connected systems ((**c**) $w3c3$ and (**d**) $w2c2$). Solid line: prey; dashed line: predator. Population sizes are depicted as per unit energy (i.e., relative to energy content in the organisms in order to normalize for the longevity of predators). Reprinted from Karsai and Kampis (2011) with permission from Elsevier

If a predator wanders to a prey-free compartment, it will starve and die there. If a prey crosses over to these fragments, then they can multiply without predation until the predators will also cross into this habitat fragment. In short, these temporal refuges make it possible that the prey builds up higher population sizes globally and this also means more food for the predators on a global scale. The predators can go extinct in some fragments, but they will thrive in others, and in this system, they can actually maintain a more stable and larger population (Fig. 3.5c, d). To understand these patterns quantitatively, we carried out extensive parameter sweeps with different predators on these habitat configurations.

The dynamics and survivability of these populations both depend on the habitat configuration and the quality of the predator species. If the predator is very efficient, it is highly probable that it drives the prey, and hence itself, to go extinct especially in a highly fragmented habitat. To test these dependencies, we applied 4 different predators to the habitat configurations studied. We assumed that if a predator gains more energy from a prey, this happens because the predator wastes less energy to catch and consume the prey (so a higher $Gain_{PD}$ means a more efficient predator).

In general, we would expect that as the predator quality increases, the number of individuals and the survival time of prey population decline, and this in turn results in also a decline of the predator population. This general pattern can be easily seen in the unfragmented case (Fig. 3.6). In an unfragmented habitat, as $Gain_{PD}$ increases, the prey population and its survival time decline, resulting in a usually small prey population also. This decrease of the prey population results in a decrease in the predator survival time if $Gain_{PD}$ is high ($Gain_{PD} > 30$). To understand the predator population's response, consider the case when $Gain_{PD}$ is high, and the predator decreases the prey population to a lower number. This also decreases the survival time of the prey, which in turn decreases the survival time of the predator again, so as a combined effect, the predator population remains near constant as a function of increasing $Gain_{PD}$ (Fig. 3.6).

Fragmentation drastically decreases the predator population size and survival time (Fig. 3.6, upper row). The increase of $Gain_{PD}$ has only a moderate effect now, the predator population rather depends on how strongly fragmented the habitat is. On the other hand, prey survival and population size increase with the fragmentation. The smaller the size of the sub-habitats, the higher the chance that predator goes extinct and if it does, the prey will flourish in the given sub-habitat.

Implementing connectors between habitat fragments results in a population boost for the predator, especially at high $Gain_{PD}$ and high fragmentation levels (Fig. 3.6, lower row). While the prey population decreases with $Gain_{PD}$ again, prey survival time radically increases when the habitat becomes more fragmented ($w >= 2$). This counter-intuitive result stems from the fact that the prey occasionally finds a temporal refuge when it escapes through the entrance to a neighboring sub-habitat that typically contains only food but no predators. These predator-free sub-habitats commonly emerge as the result of previous over-exploitation of the prey in that compartment. Because of the lack of prey, prey food has typically already completely regenerated in the area by the time the prey enters into these areas again. In this system thus a series of delays happens, which has a stabilizing

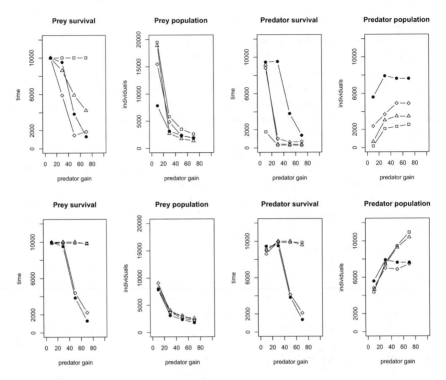

Fig. 3.6 Population size (in energy units) and survival time as a function of predator quality ($Gain_{PD}$). Upper row: The effect of fragmentation without passageways. Full dot: $w0c0$; diamond: $w1c0$; triangle: $w2c0$; square: $w3c0$. Lower row: The effect of fragmentation with connectors (corridors). Full dot: $w0c0$; diamond: $w1c1$; triangle: $w2c2$; square: $w3c3$. The figure shows mean values for 50 different runs for each parameter value. Reprinted from Karsai and Kampis (2011) with permission from Elsevier

effect. After the delayed recolonization of the prey, due to the lack of predators, an ideal condition emerges for the population growth of the prey. With typically some delay, the predators will also stumble upon the entrance points of these sub-habitats. By this point the prey has generally built up high numbers in the compartment. The predators will flourish for a while in these sub-habitats, due to the large number of prey individuals. It takes some time for the predators to find new corridors that lead to neighboring sub-habitats, where they might find a similar temporal refuge of the prey to exploit. This process ensures in general a higher predator population size combined with higher survival times in the connected fragmented habitats (Fig. 3.6, lower row).

Focusing on further analyses using effective predators ($Gain_{PD} = 50$), it becomes clear that habitat fragmentation and the reconnection of those fragments have a profound effect for both the survival and the population size of both populations (Figs. 3.7 and 3.8) When the original habitat is fragmented into just

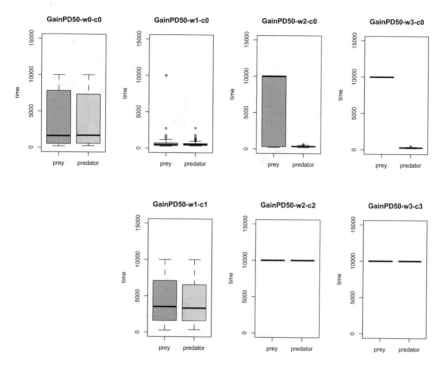

Fig. 3.7 Survival times for prey and predator populations in various habitat configurations using high quality predators ($Gain_{PD} = 50$). Upper row: No passageways ($c = 0$) (w : 0–3: number of fragmentation borders in both the horizontal and the vertical direction). Lower row: Same arrangements but with openings between the fragments ($c = w$). Reprinted from Karsai and Kampis (2011) with permission from Elsevier

a few sub-habitats (e.g., $w1c0$), the large variance in the survival time we observed in the unfragmented case disappears and the average survival time decreases for both prey and predator. This agrees with the idea of the harmful effect of habitat fragmentation. However, more fragmentation ($w2$) not only increases the survival time of the prey, but also amplifies its variability, both in survival time and numbers, due to the independent dynamics of the many isolated sub-habitats. However, this variability of the survival time disappears again when even more fragmentation is introduced ($w3$) and thus the size of the prey population increases (Fig. 3.7, top graphs). This seemingly contradictory pattern is the consequence of the results of independent dynamics in the compartments. More fragmentation means smaller compartments and as the compartment size decreases, the probability of the predator survival decreases sharply. However, in some compartments the prey will survive and these compartments will have a prey population hovering around the carrying capacity of the compartments. Connecting the sub-habitats with corridors has a positive effect on the average surviving time of both populations and in fact their survival is even better than in case of the unfragmented case (Fig. 3.7, bottom graphs). The population size of the prey increases somewhat when we implement

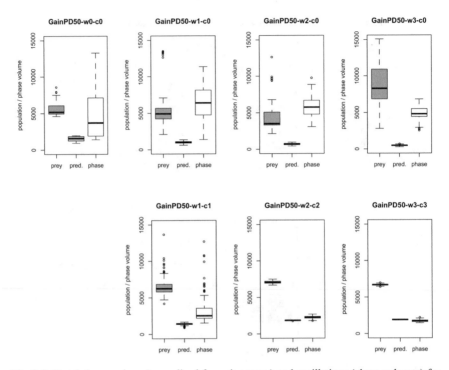

Fig. 3.8 Population numbers (normalized for unit energy) and oscillations (phase volumes) for prey and predator in various habitat configurations using high quality predators ($Gain_{PD} = 50$). Upper row: No passageways ($c = 0$) (w : 0–3: number of fragmentation borders in both the horizontal and the vertical direction). Lower row: Same arrangements but with openings between the fragments ($c = w$). Reprinted from Karsai and Kampis (2011) with permission from Elsevier

the corridors, but the most profound benefit happens to the predators, which are able to sustain a viable population in fragmented and connected habitats (Fig. 3.8, bottom graphs).

3.2.3 The Emergence of Waves in Space and Time

In an unfragmented habitat, the spatial and temporal fluctuations can be large and slow (promoted by medium values of $Gain_{PD}$). When $Gain_{PD}$ is high, the emergence of one or a few slowly moving spatial waves is frequently experienced (such waves are visible on Fig. 3.4a). In isolated fragmented habitats, fluctuations in the isolated compartments are independent, and due to the restricted smaller size of these compartments, these fluctuations tend to be fast with smaller amplitudes. The fragmented and connected habitats provide a more stable dynamics with smaller oscillations (i.e., smaller phase volumes), Fig. 3.8.

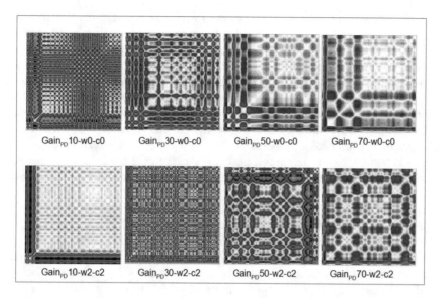

Fig. 3.9 Recurrence plots for $N = 1024$ steps of various unfragmented viz. connected fragmented systems (See Fig. 3.4 for the different habitat configurations). For high quality predators (large $Gain_{PD}$), connected fragmented systems show sustained oscillations. Reprinted from Kampis and Karsai (2011) with permission from Springer

In a connected fragmented system, an entirely new phenomenon emerges: "broken" waves appear that can percolate through boundaries via the openings and tend to form a new seed for newer waves (Fig. 3.9). To appreciate these recurrence plots, note that a perfectly regular plot (for instance, the recurrence plot of a sinus wave) shows a regular, reticular pattern. In these plots transients are visible as stripes and smudges. Disregarding transient irregularities, Fig. 3.9 shows a clear overall pattern. The upper row shows the unfragmented habitats with different predator qualities ($Gain_{PD}$). The baseline low quality predator ($Gain_{PD} = 10$) yields a quite perfect regular recurrence pattern, showing the dominance of a single frequency and its higher harmonics. Increasing $Gain_{PD}$ (going right) leads to slower and less sharp recurrences, visualized as a decay of the original regular reticular pattern. The lower row of Fig. 3.9 shows the recurrence plots of moderately fragmented connected habitats. Here, even at higher values of $Gain_{PD}$, the recurrence structure is maintained in an undamaged and sharp condition (less smudgy). We can make some analogy with what happens here to wave-breakers. Wave-breakers are obstacles that disrupt otherwise fatal sea or river waves, dispersing the effects of energy. Here, similarly, fragmentation barriers can break population waves. By breaking the otherwise fatal waves (i.e., intensive population fluctuations), the population can escape via the "leaking" of corridors into temporal refuges. Due to this, neither the prey nor the predator populations get too close to zero or can reach very high numbers. The population waves are thus damped.

This simple model system thus provides us with some new insights. The studying of habitat fragmentation without habitat destruction seems to explain the patterns found in nature. We have found fragmentation alone to be detrimental (in line with the common expectation), but if combined with connecting portals to be even advantageous, ensuring the coexistence of the prey and predator populations well above the level of the original unfragmented system. It seems that fragmentation itself is useful if it is not too extensive and fragments have links to other fragments. This commonly saves the predators from extinction and stabilizes the system. From the practical conservationist point of view, the earlier amalgamation of habitat loss and habitat fragmentation has made the evaluation of the roles of connectors such as wildlife corridors especially difficult. The debate about the effectiveness of corridors became one of the most important areas of debate for conservation biology. (Simberloff et al. 1992) asserted that the lack of data necessitates a cautious approach, in order not to invest in the construction of expensive corridors before we have quality information about their usefulness. Our study (Karsai and Kampis 2011; Kampis and Karsai 2011) showed that corridors (the connectedness of sub-habitats) indeed are essential to retain both the prey and the predator populations and ensure more stable population dynamics. Other factors such as the exact nature and properties of these corridors, differences in sub-habitat quality, and increased patchiness should be extensively studied in the future to get the full story.

References

Beier P (1993) Determining minimum habitat areas and habitat corridors for cougars. Conserv Biol 7:94–108

Caswell H, Cohen JE (1991) Disturbance, interspecific interaction and diversity in metapopulations. Biol J Linn Soc 42:193–218

Fahrig L (2003) Effects of habitat fragmentation on biodiversity. Annu Rev Ecol Syst 34:487–515

Franklin AB, Noon BR, George TL (2002) What is habitat fragmentation? Stud Avian Biol 25:20–29

Grimm V, Berger U, Bastiansen F, Eliassen S, Ginot V, Giske J, Goss-Custard J, Grand T, Heinz S, Huse G, Huth A, Jepsen JU, Jørgensen C, Mooij WM, Müller B, Pe'er G, Piou C, Railsback SF, Robbins AM, Robbins MM, Rossmanith E, Rüger N, Strand E, Souissi S, Stillman RA, Vabø R, Visser U, DeAngelis DL (2006) A standard protocol for describing individual-based and agent-based models. Ecol Model 198:115–126

Gyorffy G, Karsai I (1991) Estimation of spatio-temporal rearrangement in a patchy habitat and its application to some Auchenorrhyncha population. J Anim Ecol 60:843–855

Haddad NM, Tewksbury JJ (2006) Impacts of corridors on populations and communities. In: Crooks K, Sanjayan M (eds) Connectivity conservation. Cambridge University Press, Cambridge, pp 390–415

Hanski I (1999) Metapopulation ecology. Oxford University Press, Oxford

Harrison S, Bruna E (1999) Habitat fragmentation and large-scale conservation: what do we know for sure? Ecography 22:225–232

Kampis G, Karsai I (2011) Breaking waves in population flows. In: Kampis G, Szathmary G, Karsai I, Jordan F (eds) Advances in artificial life. Darwin Meets von Neumann. 10th European conference, ECAL 2009 (Lecture notes in artificial intelligence, vol 5777 subseries: lecture notes in computer science, vol 5778, Part II). Springer, Berlin, pp 102–109

Karsai I, Barta Z, Szilágyi G (1994) Modelling of habitat rearrangement of carabid beetles. In: Desender K, Dufrêne M, Loreau M, Luff ML, Maelfait J-P (eds) Carabid beetles: ecology and evolution. Kluwer Academic Publishers, Dordrecht, pp 153–156

Karsai I, Kampis G (2011) Connected fragmented habitats facilitate stable coexistence dynamics. Ecol Model 222:447–455

Knisley J, Schmickl T, Karsai I (2011) Compartmental Models of Migratory Dynamics. Math Model Nat Phenom 6:245–259

Levins R (1969) Some demographic and genetic consequences of environmental heterogeneity for biological control. Bull Entomol Soc Am 15:237–240

MacArthur RH, Wilson EO (1967) The Theory of Island Biogeography. Princeton University Press, Princeton

Ryall KL, Fahrig L (2006) Response of predators to loss and fragmentation of prey habitat: a review of theory. Ecology 87:1086–1093

Schweiger EW, Diffendorfer JE, Holt RD, Pierotti R, Gaines MS (2000) The interaction of habitat fragmentation. Ecol Monogr 70:383–400

Simberloff D, Farr JA, Cox J, Mehlman DW (1992) Movement Corridors: Conservation Bargains or Poor Investments? Conserv Biol 6:493–504

Chapter 4
Forest Fires: Fire Management and the Power Law

Abstract Forest fires do not only destroy plants, animals, and structures, but also change the ecology of the habitat. While research of fire dynamics is a hot topic, there are many controversial issues in the field. Forest fires are commonly considered best examples for a scale-free, power-law distribution. We developed a model of a simple ecosystem that questions this and other well-established understandings. We show that in fact forest fires are not scale-free, power-law phenomena, but are generated by different processes. Our model also shows that trees and tree-dependent animals are affected by the fires differently. In prescribed fires, the forest will burn locally with smaller fires and this will not allow for a high accumulation of fuel in the forest. While trees tend to survive or re-grow, forest animals will easily go extinct. Without prescribed fires, more fuel will accumulate, which leads to an occasional single massive fire, where a large percentage of the forest will be destroyed, but the animal population is able to rebound and spread back from unburned patches. Our model provides a clear prediction that forest animals could be endangered by prescribed fire managements.

4.1 Background

Studying forest fires initiated two lines of research, which looks very different, but finally came together and provided a better understanding of both complex systems in general and of the management of real forest fires. The research in forest fires is thus a classic example on how mathematics is able to help conservation biology and disaster management and how a real problem generates an important general understanding of complex systems.

At first glance, fire is a destructive element, which destroys the environment and resources. If we focus on naturally occurring fires only, (Feng et al. 2009) found that in the last 200 years about 6–7 million km^2 forest have been destroyed by wildfires. These wildfires do not only destroy plants, animals, people, and structures but also change the ecology of the effected habitat rapidly. The dense forest can turn into an open grassland in few months. In many ecological habitats the occurrence of fires is an active element of several processes such as ecological succession, adaptation,

© Springer Nature Switzerland AG 2020
I. Karsai et al., *Resilience and Stability of Ecological and Social Systems*,
https://doi.org/10.1007/978-3-030-54560-4_4

and the natural regeneration of the habitat (Vinton et al. 1993; Whelan 1995). An understanding of the control mechanisms of fire effects is crucial for disturbance ecology (Pascual and Guichard 2005). Understanding the underlying mechanisms provides tools for fire management, which in turn can reduce fire intensity and protect property, resources, and human life (Laverty and Williams 2000). It is also made possible, to some extent, to predict major fire events (Malamud et al. 1998) and to determine the sensitivity of fire regimes (Zinck and Grimm 2008).

The general dynamics of the forest succession results in an accumulation of dead wood material (commonly called fuel). More fuel exists in a genuine forest patch. It is more likely that fire starts or spreads in these patches and hence massive forest fires can emerge via percolation through these patches. The early method to stop forest fires focused on stopping the burns at the beginning, but the fire exclusion in these forests resulted in dense canopies with densely packed high fuel patches (Brown 1985; Ferry et al. 1995). These management practices resulted in rarer, but in turn very massive and sometimes even unmanageable forest fires. Different methods have been developed to assess and quantify the fire hazard based on measuring canopies, fuel quantities, and similar variables (Finney 2005; Bajocco et al. 2009; Keane et al. 2010). These assessments made it possible to practice prescribed forest fires, where the forest management actively and in a controlled manner burns part of the forest to decrease tree density and eliminate fuel accumulation. This practice is commonly heralded as a solution for protecting forests from massive destructions and for increasing the diversity of organisms in these forests. The effect of fire on vegetation has been extensively studied, and also studies on the effect of fire on animals have provided essential insights into the complex network of causality in ecosystems (Fons et al. 1993; Pons et al. 2003; Zamora et al. 2010).

Conservation biology commonly focuses on plants and birds and in these organisms an increase of diversity could be observed, but the less mobile and often more specialized animals such as insects can actually be endangered by the prescribed fire management practice (Horton and Mannan 1988; Russell et al. 1999; Harper et al. 2000). One of the goals of our model is to demonstrate such a case.

Theoretical studies had interesting findings and have uncovered that fire sequences can be considered as fractal processes with a high degree of time clustering of events. These models are generally based on the mechanisms of self-organized criticality, assuming to have a slow driving energy input and rare avalanche-like dissipation events that by contrast have a more rapid dynamics (Song et al. 2001). The accumulation of fuel (such as the emergence of more and more dead branches) in the aging forest is the slow driving energy input (Zinck and Grimm 2008) and the avalanche-like dissipation event is the burning of this fuel in a short time. The forest fire, similar to the famous sand pile avalanches, (Bak 1996) has been an example of the critical scaling behavior in "turbulent" nonequilibrium systems (Bak et al. 1990). The early model was further elaborated (Drossel and Schwabl 1992; Zinck and Grimm 2008) and the authors have also introduced a "lightning parameter" to initialize fires by direct control. Niazi et al. (2010) also developed a realistic, verified, and validated agent-based forest fire simulation (and also provided an overview of forest fire simulation models).

One of the cornerstones of these works was an assumed power-law distribution of the forest fires. Considering forest fires as scale invariant universal natural phenomena, similar to the scaling laws in biology, has been a very active field of research. However, self-organized criticality is not the only possible mechanism for generating power-law distributions in natural phenomena (Solow 2005). For example, power-law relationships have been found between the frequency of fires and the size of the burned areas (Malamud et al. 1998; Ricotta et al. 1999, 2001; Song et al. 2001; Telesca and Lasaponara 2010).

While theoretical works focused on the power-law distributions and on the possible mechanisms able to generate such distributions, data on real fires have shown a significant divergence from the theory. For example, analyzing wildfires on US federal lands between 1986 and 1996 showed that while the distribution of fires spans 6 orders of magnitude, the distribution at its tail shows an exponential cutoff rather than a power function (Newman 2005). Ricotta et al. (1999) analyzed the distribution of a large number of fires in Italy and concluded that self-organized criticality could be a good explanation of the power-law distribution found, however, the authors used only the middle part of the distribution, and the first and last half of their data does not even look close to a power-law distribution. These and similar problems fostered the development of new analytical methods and statistical tests to evaluate data and to scrutinize curve fitting methods (we provide some pointers to the latter at the model description section).

4.2   An Individual- (Agent) Based Model of Forest Fire

4.2.1   Description of the Model

The purpose of our model is to understand how forest fires affect trees and forest animals and also to provide simulation data for testing the power-law (PL) hypothesis in the ecosystem. Our model is essentially different from the classical cellular automata approach of Bak et al. (1990) or Drossel and Schwabl (1992). In our model, namely, fire sprites are defined, which can reproduce and move randomly, therefore a neighboring tree has a chance to avoid burning down. The basic mechanism of spreading has been earlier identified as a percolation process (see Von Niessen and Blumen 1986; Henley 1989; Beer and Enting 1990), which we wanted to indeed re-invigorate in our model using the agent-based approach.

In developing the present model, we had multiple goals: (1) keeping the fire process minimalistic, yet using a fuel-based mechanism based on a simple percolation system; (2) assessing the effect of fire on both the trees and on animal populations (the latter is of course dependent on the former); (3) providing data for analyzing the distribution of the sizes of the fires.

The model was developed in the Netlogo 5.3.1 environment and is described in full details, using the ODD protocol (Grimm et al. 2006), in (Karsai et al. 2016). The model is also deposited in the Netlogo community site[1] and its companion site.[2] Here we describe the core of the model in non-technical terms.

The model has three hierarchical levels: entities (trees, animals, and fires), interactions, and environment. Entities (agents) have a rule set and they form populations (or collectives, Fig. 4.1). The populations are characterized by the census of each organism type at the end of a given year. The number of burned trees and animals are also counted. The agents have a very minimalistic rule set. Trees and fires are characterized by their position and can be alive or dead. Animals are characterized, besides the same, by two further internal state variables: the speed of their movement, as well as the current activity performed (which can be moving or breeding). Time progresses in ticks where a single tick represents a month and hence 12 ticks a year. Animals and fires disperse at each tick while other events such as tree and animal reproduction as well as the ignition of the fires (i.e., fire initiation) happen yearly. The time and space scales chosen allow for a study of a large-scale dynamics, while the simulations still run quick enough for many parallel runs making extensive parameter sweeps possible.

The model does not intend to grasp characteristics of a specific system or has parameters estimated from natural systems directly. The selection of the parameter values was motivated by the desire to have sufficiently many interactions between the different agents in a 100-year period, so that we do not need very long runs and a high number of replicates. The area selected is large enough to allow for a diversification by local events. This ensures a delay effect: if the trees are burned down in an area, the whole animal population cannot simply escape into a better habitat but many individuals tend to die. The model is otherwise highly abstract. For simplicity, organisms do not even have a metabolism, but they can reproduce and die naturally and by the fire accidentally. The mortality and reproduction of animals are also dependent on the number of healthy trees in their immediate vicinity. This important assumption explicitly wanted to capture the effect of fires for animals that are highly dependent on the forest and cannot simply evade the fires by escaping.

Fires are initiated in a random position every year, so that each time a random integer between 0 and N_f (a fire initiation parameter) is picked. The corresponding number of fire seeds (fire seed agents) are introduced in the system. These can be conceived as sparks or lightning hits, and only if they directly hit a tree do they become a fire sprite. The fire can spread if there is a tree (i.e., fuel) next to a fire sprite. In this case the fire sprite will produce an offspring and both the offspring and the "mother sprite" will move further in random directions. A fire is extinguished if there are no trees in its neighborhood. For simplicity we did not implement the effects of wind or long-distance fire spreading (i.e., jumping).

---

[1] http://ccl.northwestern.edu/netlogo/models/community/Fire%20in%20the%20forest.

[2] https://sites.google.com/site/springerbook2020/chapter-4.

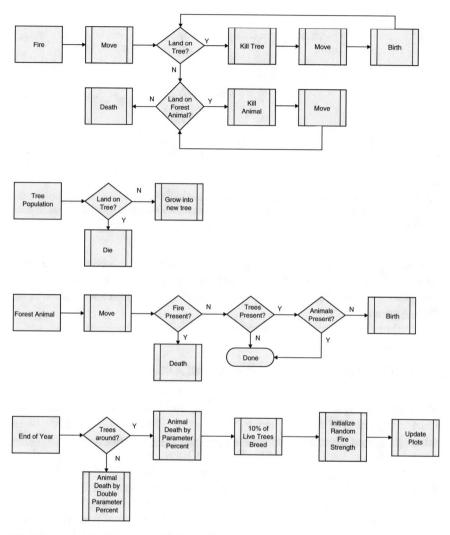

Fig. 4.1 Process overview of the model. Each agent has a set of rules to carry out (top 3 flowcharts); the timing of events is shown on the bottom chart. Reprinted from Karsai et al. (2016) with permission from Elsevier

The population's dynamics and individual behaviors emerge from interactions at the individual (agent) level. Adaptation and fitness were not explicitly sought or included in this model, yet some such features were reflected in the rules selected for the individuals (for example, the effect that crowding prohibits reproduction). All sensing and interaction are made strictly local to the agents. This means that individuals "know" (or access) their own current activity status and they can check (i.e., sense) the existence and the status of other agents in their neighborhood—such as recognizing conspecifics or the vicinity of trees. A set of preliminary

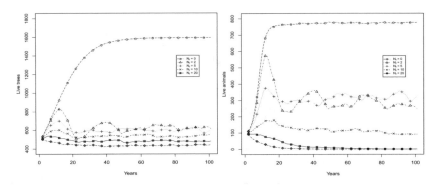

Fig. 4.2 The average number of trees (left) and forest animals (right) in the studied habitat over 100 years, in the function of fire parameter N_f. Reprinted from Karsai et al. (2016) with permission from Elsevier

experiments has been performed to test the reliability of the model for example, the emergence of carrying capacity without explicitly implementing it or the emergence of diminishing breeding with a growing population size (Karsai et al. 2016, Appendix).

4.2.2   The Effect of Fire on Plants and Animals

When fires are introduced in the model habitat, the average tree population drops about to one fourth of the maximum value observed (1600 trees). The animals seem to have a stronger dependency on the fire parameter N_f (the number of fire initiation events/year) than trees do (Fig. 4.2). If the fire parameter N_f is small to medium (i.e., Nf is between 2−10), then the animal population is typically sustainable, but at higher N_f values the animals will became extinct from the habitat. Trees will survive even high frequency fires. On the average, with increasing fire frequency N_f, both the trees and the animals show decreasing population numbers (Mann–Whitney U test for pairwise comparisons, $p < 0.05$, $N = 100$); except that, counter-intuitively, significantly more animals exist in the habitats at a medium value of the fire parameter ($N_f = 5$) than in the case of weak fires ($N_f = 2$; Mann–Whitney U test, $p < 0.05$, $N = 100$; Fig. 4.3).

The counter-intuitive case about animal numbers, that $N_f = 5$ produced a larger living animal population than did habitats with $N_f = 2$, is in fact very interesting. We should note that the fact does not mean that in the habitats with $N_f = 5$ fewer animals would be killed. Quite the opposite, a lot more animals were burned in these cases than in any other (Fig. 4.3; Mann–Whitney U test for pairwise comparisons $p < 0.05$, $N = 100$), therefore the difference described above is not due to a decreased mortality. This pattern is similar to the maximum sustainable yield phenomenon, where populations about half the size of their carrying capacity

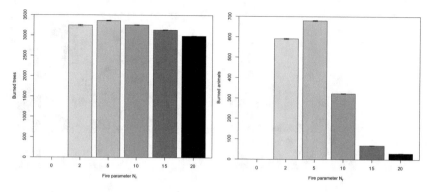

Fig. 4.3 The total value of burned trees (left) and burned forest animals (right) in the studied habitat as a function of the fire parameter (maximum number of fire initiation points/year). The 50 parallel simulations of each year from year 1 to year 100 were averaged, and these yearly average values were totaled. We found significant differences between the yearly burned animals with different fire parameters, and the yearly burned trees showed a significant difference between fire parameter $N_f = 20$ vs. 10, and $N_f = 20$ vs. 5 but not when N_f was between 1 and 15 (Mann–Whitney U test $p < 0.05$, $N = 200$). Reprinted from Karsai et al. (2016) with permission from Elsevier

are able to produce the largest number of offspring. This automatically emerged in our model at moderate values of the fire parameter N_f. Because our model is an agent based model (ABM), the carrying capacity and the maximum sustainable yield phenomenon were not explicitly built into the model, but were found to be emergent properties of the system.

It is also clear that when the fire parameter N_f is very large, not many animals can be burned anymore, because the animal population size is much smaller (they cannot regenerate much before a new fire effect hits them). The number of burned trees were between 3000 and 3500 with a smaller dependency on the fire parameter (Fig. 4.3).

On the average, with the increase of the fire parameter, more "live" fires (the number of existing fire sprites at the end of the year) and "dead" (i.e., extinguished) fire sprites can be observed (Fig. 4.4). Hence the patterns we observed in the animals and in the trees (Figs. 4.4 and 4.5) cannot be simply derived from the average number of fire sprites directly. High values of the fire parameter N_f (15–20) resulted in 30–40 extinguished fire sprites with 8–10 still alive yearly on average, and this pattern is very similar from year to year (with moderate fluctuations, Fig. 4.5). On the other hand, at low fire parameter values, the fire initialization (lightning) commonly missed to ignite the trees, thus causing no fires at all in the given year (and this could be repeated in several consecutive years). Over these fire-free years, both the trees and animals are freely multiplying and this can build up a large quantity of fuel (i.e., trees to burn). As the fuel quantity is increasing in the habitat, the successful fire initiation also increases (there will be less possibility to miss). This in turn commonly generates a very extensive burning, decreasing both the tree and the animal populations significantly. While trees could regenerate after such an

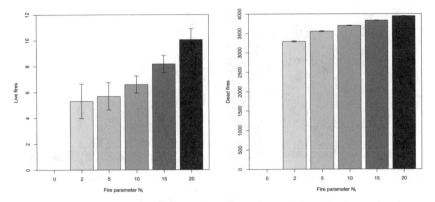

Fig. 4.4 The average number of live fire sprites at the end of the year and the total number of dead (extinguished) fire sprites in the studied habitat as a function of the fire parameter N_f. Fire sprites alive at the end of the year from year 50 to year 100 were averaged and the yearly average values were used to calculate mean and standard deviations. Similarly, extinguished fire sprites were counted, averaged by years and added up for 100 years. We found significant differences between the numbers of fire sprites in habitats with different fire parameters N_f (Mann–Whitney U test $p < 0.05$, $N = 100$ for live and $N = 200$ for dead fire sprites). Reprinted from Karsai et al. (2016) with permission from Elsevier

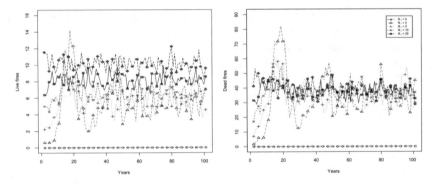

Fig. 4.5 The average number of live and dead fire sprites in the studied habitat in 100 years as the function of fire parameter (maximum number of fire initiation points/year). Reprinted from Karsai et al. (2016) with permission from Elsevier

event quickly from their seeds, animals also suffer a secondary detrimental effect by losing the forested area where they could breed and survive. After a massive burn, the habitat will lose most of its fuel (trees) and therefore in the consecutive years in general the lightning will miss, and no fire will be initiated for several years again. This dynamics results in a fluctuating dynamics (Fig. 4.5) and a notable distribution (which we will study next).

4.3   Power-Law Analysis of the Distribution of Fires

4.3.1   Methods

Due to equivocal results in the literature, we also wanted to study the fire size distributions in the simulated ecosystem. We wanted to understand how the distribution depends on the main control parameter N_f that describes the upper limit of the random fire seeds ignited in every year. Simple curve fitting to a power function generally did not produce a very good fit, therefore, as described above, it was a common practice earlier to leave out a part of the data from the analysis and to focus only on those data points that fit best for the power function. This is not only a questionable practice, but as we will see, it actually misses the most important message from the data, namely that the fires have a complex distribution and actually do not follow a power law. To study this, we used a more complicated and we think more correct approach, which we describe here shortly.

For the statistical analyses, data from the parallel simulations were pooled. A log-log cumulative distribution was then fitted to different theoretical distributions using a maximum likelihood method (Newman 2005) and tested by the Vuong test (Vuong 1989). Afterwards, we have used advanced tools for testing distributions for the power-law property introduced by Clauset et al. (2009). The authors show that in power-law testing, instead of performing a simple curve fitting, a complex algorithm built on the cumulative distribution function (CDF) should be applied. This procedure is unfortunately not generally known, albeit it replaces the older methods of distribution fitting and provides more reliable results.

The starting observation of the approach is that typical empirical or model-generated distributions are produced by mixtures of different mechanisms, not necessarily a single one. For example, real systems always have limiting factors, therefore their empirical distributions tend to behave differently at very small and very large values. The process of the power-law fitting should thus include the identification of an inevitable cutoff (denoted X_{min} later). The procedure itself relies on a recursive (bootstrap) use of the maximum likelihood method to estimate a cutoff value, above which the power-law property can be tested at all. If the cutoff value is zero, this means that the entire dataset comes from a single distribution. The Vuong test (Vuong 1989) has been specifically designed for comparing different distributions and it can be applied to test the distribution both before and after the cutoff point (X_{min}). All these methods have been implemented together, for example, in the R package *poweRlaw* (Gillespie 2015) which we actually used.

4.3.2   Size Distribution of Fires in the Habitat

During the 100 years studied and in the 50 parallel runs taken, around 200,000 fires have been burning in our modeled habitat. We pooled the data belonging to the same

Fig. 4.6 Cumulative
distribution of fires (sample
plot for $N_f = 2$) in our
simulated habitat. On the x
axis, the total number of fires
(dead + live) are shown in a
given run. On the y axis,
cumulative frequency: that is,
y_{max} is the total number of
fires in 50 runs. Reprinted
from Karsai et al. (2016) with
permission from Elsevier

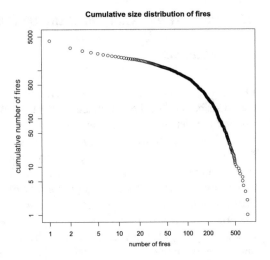

N_f (fire initiation parameter) values to study the effect of fire initiation frequency
on the fire distributions. The cumulative distribution functions of fire sizes in our
model show a slow decrease at first and then an accelerated drop (Fig. 4.6 shows
an example). The distributions seemed to have two visually separable parts, and
thus we explicitly tested if they come from a single distribution or if we have two
different distributions with a cutoff point. If there exists a cutoff point, then two
different behaviors (and probably two distributions) are experienced for small fires
and large fires, respectively. The testable question here is whether these parts indeed
come from different distributions, or from an identical distribution with different
parameters? Either way, our question was: is there a power-law behavior shown by
the results (in the whole distribution or in one of the distribution segments)? To test
these issues, first we estimated the cutoff values (X_{min}) above which the power-law
property could be tested at all, then we have fitted both the resulting segments to
power-law, lognormal, and exponential distributions (Table 4.1).

　　From the results we can conclude that the power-law hypothesis can be safely
rejected. The PL distribution occurred only for one segment of the 10, and there
only as an alternative, possible explanation (in $N_f = 10$, second segment). In
all other cases, different distributions were found (Table 4.1). This means that
there is not even a single domain of N_f in which power law would be the only
available explanation of the observed data distribution. Instead, the lognormal and
the exponential distributions dominate for both "small fires" and "large files" for all
values of N_f tested. Our results also indicate that the essential mechanism for the
two domains found may in fact be identical, only with different parameters (and also
given that EXP is effectively a subcase of LN, as the exponential distribution can be
obtained from the lognormal). So we have also tested if a single EXP or LN with
the *same* parameters for the whole interval could fit the entire distributions—but this
hypothesis was rejected. This indeed shows that if a single mechanism generates the
different distribution segments, it does so under different parameters.

Table 4.1 Characteristic distributions at various values of N_f and their quantitative relations. Items marked as [a] are at $0.05 < p < 0.08$ (i.e., "marginally significant"), otherwise always $p < 0.05$

Fire parameter N_f	2
Small fires	LN
Large fires	EXP[a]
X_{min}	274
Sum of small fires	145,441
Sum of large fires	45,800
Ratio of small/large	3.18
Fire parameter N_f	**5**
Small fires	EXP
Large fires	EXP/LN
X_{min}	102
Sum of small fires	136,007
Sum of large fires	69,890
Ratio of small/large	1.95
Fire parameter N_f	**10**
Small fires	EXP
Large fires	EXP/LN/PL
X_{min}	121
Sum of small fires	189,963
Sum of large fires	27,964
Ratio of small/large	6.79
Fire parameter N_f	**15**
Small fires	LN
Large fires	EXP[a]
X_{min}	104
Sum of small fires	199,447
Sum of large fires	33,023
Ratio of small/large	6.04
Fire parameter N_f	**20**
Small fires	LN
Large fires	EXP
X_{min}	98
Sum of small fires	209,492
Sum of large fires	38,375
Ratio of small/large	5.46

We can also observe that whereas the total number of fires grows essentially linearly with the increasing number of maximum fire seeds N_f, at the same time the ratio of small to large fires changes differently (Fig. 4.7). At small-medium values ($N_f = 2$–5) the ratios are about half of what we can observe later, indicating a relatively higher number of massive fires. At $N_f = 10$ there is a sudden peak indicating the highest relative number of small fires compared to the large ones, and with increasing N_f this ratio will decrease only moderately.

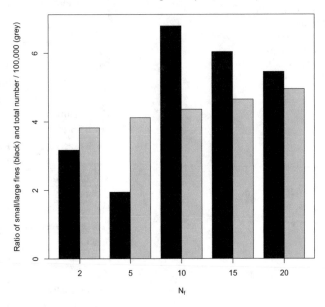

Fig. 4.7 Ratio of small/large fires (black) and the total number of fires (gray, scaled down by a factor of 100,000 for plotting). Reprinted from Karsai et al. (2016) with permission from Elsevier

4.3.3 Practical and Theoretical Conclusions

Schenk et al. (2000) suggested that in the self-organized criticality forest fire model, in fact, two qualitatively different types of fires occur. Type one fires can burn down a patch of high tree density and it can be characterized with a higher fractal dimension, whereas type two fires can burn down a tree cluster and can be characterized by a smaller fractal dimension (within a region of a tree density below the percolation threshold). This is an interesting insight and while we did not estimate the embedded fractal dimensions, our model indicates similar patterns and our analysis shows that the scaling behavior of the system cannot be characterized by only one single length scale. The superposition of the two types of fires creates a more complex picture than a simple power law suggested by earlier works.

Although our model is quite simple and our assumptions certainly contain some strong simplifications, it should be noted that the cumulative distributions of fire in our simulations are very similar than those of natural fires (Newman 2005; Ricotta et al. 1999). Our model can be augmented by further factors. For example, our model is based on the general pattern that a higher fuel amount generates more severe fires (Dodge 1972), but this pattern in nature may be overridden by the effect of fuel moisture, weather conditions, wind, fuel type, and other factors (Turner and Romme 1994; Hely et al. 2001; Behm et al. 2004). Yet even without these, our model showed a cutoff and a natural-like cumulative size distribution, therefore this seems to be an inherent property of these systems and its dependence from the mentioned facts could be neglected. For example, both in real-world and model fires, it seems we can observe an emergence of some kind of upper bound depending on the size of the fuel bed and the connectedness of the fuel cells.

About the importance of theoretical and abstract models in general, we completely agree with Busing and Mailly (2004) saying that

> Models are particularly useful in pointing out weaknesses in our understanding of ecological systems and in pointing out which system components are most important; they provide new perspectives and insights that may lead to the development of testable hypotheses.

In fact, our model suggests practical insights on fire management and conservation biology. The now-common practice of prescribed burning seems to be a great solution for protecting the forest of massive fires and increases the diversity of organisms. However, common surveys on animal diversity are focusing largely on plants and birds, and these organisms are the quickest re-colonizers. Our simulation shows a different picture. We presented a model in which trees (plants) and tree-dependent animals with limited movements (such as many invertebrates, amphibians in reality) are affected by the fires differently. When there is a high probability of fire initiation (such as the case in prescribed fires), the forest will burn locally with smaller fires and this will not allow a high accumulation of fuel in the forest (DeBano et al. 1998). Our model shows that while most trees will survive or regrow, in such a case many forest animals will easily become extinct. When the fire initiation probability is small, the occurrence probability of fires is smaller, however more fuel will accumulate in time, and all this leads to an occasional massive fire, where a large percentage of the forest will be destroyed. While these occasional big fires are also devastating to the animal population, they are able to rebound and slowly spread back from the unburned patches. We suggest that prescribed fires need to be evaluated based on the whole ecosystem and especially on their most vulnerable populations. Our model predicts that animal populations that are strongly connected to forest plants are, due to their limited movement ability, resilient to fires that happen rarely even if some of these are extremely extensive, but they tolerate much less the occurrence of frequent fires. Our model provides a clear prediction that forest animals could be endangered by prescribed fire managements (see also Horton and Mannan 1988; Russell et al. 1999; Harper et al. 2000).

References

Bajocco S, Rosati L, Ricotta C (2009) Knowing fire incidence through fuel phenology: a remotely sensed approach. Ecol Model 221:59–66

Bak P (1996) How nature works: the science of self-organized criticality. Nature 383(6603):772–773

Bak P, Chen K, Tang Ch (1990) A forest-fire model and some thoughts on turbulence. Phys Lett A 147(5):297–300

Beer T, Enting IG (1990) Fire spread and percolation modelling. Math Comput Model 13(11):77–96

Behm AL, Duryea ML, Long AJ, Zipperer WC (2004) Flammability of native understory species in pine flatwood and hardwood hammock ecosystems and implications for the wildland-urban interface. Int J Wildland Fire 13:355–365

Brown JK (1985) The "unnatural fuel buildup" issue. In: Lotan JE, Kilgore BM, Fischer WC, Mutch RW (eds) Symposium and workshop on wilderness fire. U.S. Department of Agriculture, Forest Service, Intermountain Forest and Range Experiment Station, Missoula, pp 127–128

Busing RT, Mailly D (2004) Advances in spatial, individual-based modelling of forest dynamics. J Veg Sci 15:831–842

Clauset A, Shalizi CR, Newman MEJ (2009) Power-law distributions in empirical data. SIAM Rev 51(4):661–703

DeBano LF, Neary DG, Folliot PF (1998) Fire's effects on ecosystems. John Wiley and Sons, New York

Dodge M (1972) Forest fuel accumulation—a growing problem. Science 177:139–142

Drossel B, Schwabl F (1992) Self-organized critical forest-fire model. Phys Rev Lett 69:1629–1992

Feng G, Yan X, Hu X (2009) State estimation using particle filters in wildfire spread simulation. In: Proceedings of the 2009 spring simulation multiconference (SpringSim '09). Society for Computer Simulation International, San Diego, p 8

Ferry GW, Clark RG, Montgomery RE, Mutch RW, Leenhouts WP, Zimmerman GT (1995) Altered fire regimes within fire-adapted ecosystems. U.S Department of the Interior, National Biological Service, Washington

Finney MA (2005) The challenge of quantitative risk analysis for wildland fire. For Ecol Manag 211:97–108

Fons R, Grabulosa I, Feliu C, Mas-Coma S, Galan-Puchades MT, Comes A (1993) Post-fire dynamics of a small mammal community in Mediterranean forest (Quercus suber). In: Trabaud L, Prodon R (eds) Fire in Mediterranean ecosystems. ECSC-EEC-EAEC, Brussels-Luxembourg, pp 259–270

Gillespie CS (2015) Fitting heavy tailed distributions: the poweRlaw package. J Stat Softw 64(2):1–16. (Also see vignettes of the R package poweRlaw by the same author: "The poweRlaw package: a general overview", "The poweRlaw package: Comparing distributions", "The poweRlaw package: Examples". https://cran.r-project.org/web/packages/poweRlaw/vignettes/)

Grimm V, Berger U, Bastiansen F, Eliassen S, Ginot V, Giske J, Goss-Custard J, Grand T, Heinz S, Huse G, Huth A, Jepsen JU, Jørgensen C, Mooij WM, Müller B, Pe'er G, Piou C, Railsback SF, Robbins AM, Robbins MM, Rossmanith E, Rüger N, Strand E, Souissi S, Stillman RA, Vabø R, Visser U, DeAngelis DL (2006) A standard protocol for describing individual-based and agent-based models. Ecol Model 198:115–126

Harper MG, Dietrich CH, Larimore RL, Tessene PA (2000) Effects of prescribed fire on prairie arthropods: an enclosure study. Nat Areas J 20(4):325–335

Hely C, Flannigan M, Bergeron Y, McRae D (2001) Role of vegetation and weather on fire behavior in the Canadian mixed wood boreal forest using two fire behavior prediction systems. Can J For Res 31:430–441

Henley CL (1989) Self-organized percolation: a simpler model. Bull Am Phys Soc 34:838

Horton SP, Mannan RW (1988) Effects of prescribed fire on snags and cavity nesting birds in southeastern Arizona pine forests. Wildl Soc Bull 37–44

Karsai I, Roland B, Kampis G (2016) The effect of fire on an abstract forest ecosystem: an agent based study. Ecol Complexity 28:12–23

Keane RE, Drury SA, Karau EC, Hessburg PF, Reynolds KM (2010) A method for mapping fire hazard and risk across multiple scales and its application in fire management. Ecol Model 221(1):2–18

Laverty L, Williams J (2000) Protecting people and sustaining resources in fire adapted ecosystems: a cohesive strategy. Forest Service response to GAO Report. GAO/RCED 99–65. USDA Forest Service, Washington

Malamud BD, Morein G, Turcotte DL (1998) Forest fires: an example of self-organized critical behavior. Science 281:1840–1841

Newman MEJ (2005) Power laws, Pareto distributions and Zipf's law. Contemp Phys 46(5):323–351

Niazi MA, Siddique Q, Hussain A, Kolberg M (2010) Verification and validation of an agent-based forest fire simulation model. In: Proceedings of the 2010 spring simulation multiconference (SpringSim '10). Society for Computer Simulation International, San Diego, pp 1–8

Pascual M, Guichard F (2005) Criticality and disturbance in spatial ecological systems. Trends Ecol Evol 20:88–95

Pons P, Henry PY, Gargallo G, Prodon R, Lebreton JD (2003) Local survival after fire in Mediterranean shrublands: combining capture-recapture data over several bird species. Popul Ecol 45:187–196

Ricotta C, Avena G, Marchetti M (1999) The flaming sandpile: self-organized criticality and wildfires. Ecol Model 119:73–77

Ricotta C, Arianoutsou M, Díaz-Delgado R, Duguy B, Lloret F, Maroudi E, Mazzoleni S, Moreno JM, Rambal S, Vallejo R, Vázquez A (2001) Selforganized criticality of wildfires ecologically revisited. Ecol Model 141(1):307–311

Russell KR, Van Lear DH, Guynn DC (1999) Prescribed fire effects on herpetofauna: review and management implications. Wildl Soc Bull 374–384

Schenk K, Drossel B, Clar S, Schwabl F (2000) Finite-size effects in the selforganized critical forest-fire model. Eur Phys J 15:177–185

Solow AR (2005) Power laws without complexity. Ecol Lett 8:361–363

Song W, Weicheng F, Binghong W, Jianjun Z (2001) Self-organized criticality of forest fire in China. Ecol Model 145:61–68

Telesca L, Lasaponara R (2010) Analysis of time-scaling properties in forest-fire sequence observed in Italy. Ecol Model 221:90–93

Turner MG, Romme WH (1994) Landscape dynamics in crown fire ecosystems. Landsc Ecol 9: 59–77

Vinton MA, Hartnett D, Finck EJ, Briggs JM (1993) Interactive effects of fire, bison grazing and plant community composition in Tallgrass Prairie. Am Midl Nat 129:10–18

Von Niessen W, Blumen A (1986) Dynamics of forest fires as a directed percolation model. J Phys A: Math Gen 19(5): L289

Vuong QH (1989) Likelihood ratio tests for model selection and non-nested hypotheses. Econometrica 57(2):307–333

Whelan RJ (1995) The ecology of fire. Cambridge University Press, Cambridge

Zamora R, Molina-Martínez JR, Herrera MA, Rodríguez y Silva F (2010) A model for wildfire prevention planning in game resources. Ecol Model 221(1):19–26

Zinck RD, Grimm V (2008) More realistic than anticipated: a classical forest-fire model from statistical physics captures real fire shapes. Open Ecol J 1:8–13

Chapter 5
Material Flow, Task Partition, and Self-Organization in Wasp Societies

Abstract Insect societies are a prime model system to investigate the processes of homeostasis, self-organization and the emergent properties of complex systems. Using both top-down and bottom-up modeling techniques, we show here how effective a simple regulatory mechanism (we name it "common stomach regulation") can be in sustaining the stability of the system in these societies. Wasp societies are resilient to external perturbations and they quickly adjust their workforce to compensate or to establish new equilibria. Flexibility of behavior at the individual level and self-organized feedback mechanisms at the system level are the key for both for this resilience and the large-scale internal development that colonies often undergo. Using meta-analyses, we also showed how point attractors are replaced with oscillations in these systems, where flexibility at the individual level has a cost. The emergence of specialists is not a prerequisite, but rather a consequence in these systems. In short, specialists emerge because larger systems have a relatively smaller variation, hence a smoother functioning. These systems are also not scaling linearly, for example, the number of foragers is kept very low even in large colonies, which is a mechanism that has probably evolved against predators.

5.1 Background

5.1.1 Task Partitioning

Living systems are open dissipative systems where energy and material is flowing through the system and it is these flows that maintain the highly organized complex nature of organisms. A homeostatic self-regulation is fundamental for many systems besides biological ones, including physical, chemical, societal, and even economical systems. These systems are commonly in the center of scientific inquiry due to their fundamental effects on our life, as climate change exemplifies. Many of these systems are extremely complicated with a high number of elements and a staggering complexity of interaction networks. Often, however, even the elementary relationships between units or the units themselves are unknown. By contrast, we have numerous observations and experimental data on social insects, especially

© Springer Nature Switzerland AG 2020

I. Karsai et al., *Resilience and Stability of Ecological and Social Systems*,
https://doi.org/10.1007/978-3-030-54560-4_5

on honeybees and economically important ants. This accessibility is one of the main reasons that research on insect societies became a pioneer model system to investigate the processes of homeostasis, self-organization and the emergent properties of complex systems (Schmickl and Karsai 2018).

Insect societies provide an excellent example of de-centralized self-organized systems. The colony generally consists of a queen, which is not a central organizer of the colony as one might believe, but just an egg-laying machine. While her interactions with other individuals can be directly and chemically important, but they are also quite limited. The queen does not work as a leader or foreman. The large workforce in the colony is entirely self-organized, via several mechanisms that are generally based on simple cues and local interactions with other workers and with the objects of the work (Karsai 1999). Work in the colony is carried out by parallel processing. Generally, parallel processing requires the existence of several agents or units, plus mechanisms that ensure the specialization and organization of these units into a complex and efficient system. In social insects, the colony conducts all of its operations concurrently, with many individuals employed on the same job. Reliability theory posits that redundancy at this subunit level is more efficient than redundancy at the system level (Barlow and Proschan 1975).

Parallel processing happens via division of labor and task partitioning. Although recently there are many debates on the meaning of these concepts, in this work we simply focus on a special type of worker force allocation (we will call it task partitioning nevertheless). It is common for insect societies that a work requires several steps to finish. For example, adding building material into the nest (construction task) requires first to leave the nest and to forage for water. When enough water is collected, the given wasp will fly to a place where there is building material available (i.e., old dead wood) and here the wasp extracts fibers and mixes them with the water she has carried. When this pulp is processed, the wasp will fly back to the nest and commence the construction. Obviously, this is a long sequence and the wasp needs to do it in the specified order, and also, she needs to learn the position of water and pulp sources and needs to avoid predators. A tall order! *Polistes* and other wasp species use this type of sequential organization of work for building (Karsai et al. 1996; Karsai and Pénzes 1993, 1998). Societies with a larger number of workers can partition this task into subtasks and task-specific specialists can emerge, which can be more effective than the "jack of all trades" *Polistes* workers are.

Task partitioning relates to the colony size. Colony size determines the size of the workforce. Whether few generalists or a set of parallel processing specialists are more effective depends on the size of the workforce. A single wasp needs to carry out the steps in a sequential order, but a small colony can partition the work into 2 subtasks (Fig. 5.1a vs. b and c). In these colonies, the water foragers fly out to the water source and collect water. Instead of going to the pulp source, these wasps return home and download the water to other wasps. These receiver wasps will become the pulp foragers that also double as builders. After receiving the water, these individuals go to the pulp collecting place and collect pulp, then they return to the nest where they build this pulp into the nest. A problem for this system can occur

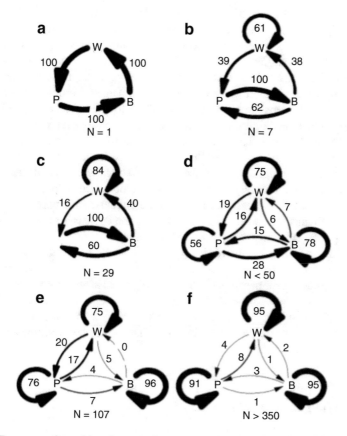

Fig. 5.1 Frequency of transitions between three construction tasks, building (B), pulp foraging (P), and water foraging (W). (**a**) Single foundress, as in *Polistes* (**b**) *Vespula sylvestris*, colony of seven individuals, 44 transitions of one individual, from ref. 28. (**c**) *Polistes fuscatus*, colony of 29 individuals, 155 transitions, recalculated from ref. 29. (**d**) *Polybia occidentalis*, data pooled from four colonies of 50 individuals, 797 transitions (15). (**e**) *Metapolybia mesoamerica* (30), colony of 107 individuals, 117 transitions (this study). (**f**) *Polybia occidentalis*, data pooled from three colonies, larger than 350 individuals, 2085 transitions, from ref. 15. Width of arrows corresponds to frequency; numerals indicate exact values. In (**b**) and (**c**), every pulp forager also built with her pulp (regardless of sharing) as indicated by the large, straight arrows. (Reprinted from Karsai and Wenzel 1998 with permission from National Academy of Sciences)

if the 2 tasks are unbalanced. For instance, if there are too many water foragers and not enough pulp collectors, then the water collectors need to wait for a long time to download their water (queuing delay). Individuals that are waiting for a long time to finish their job will probably abandon it and switch to some other job instead. In such a case the switching and waiting can make this system less productive than applying "jack of all trades" workers. This can happen especially in small colonies.

In *Metapolybia* and *Polybia* wasps, which have large colonies and breed by colony swarms, the construction is partitioned into water foraging, pulp foraging and building subtasks and the individuals who engage in one task rarely switch to another (Fig. 5.1d–f). These wasps also have two further innovations evolved. The water foragers do not give water directly to the pulp foragers, but download it to the wasps that are sitting on the nest. The pulp foragers also collect water from these individuals instead of looking for an entering water forager. These sitting wasps comprise a common water tank, which we named "common stomach," and it considerably shortens the queuing delays of the foragers. The other evolutionary innovation is that the pulp foragers can process and deliver a large quantity of pulp at the same time, because they can give all that away to numerous builder wasps. For the smaller colonies, the amount of the pulp delivered to the nest needed to be small enough, so that a pulp forager could also be able to build with the collected material. In the *Metaplybia* and *Polybia* wasp colonies the delivered pulp is macerated into small chunks on the nest and distributed to the builders, which are able to handle these small chunks for construction. This innovation decreases the flight frequency of the pulp foragers almost by a degree of magnitude.

There are a series of interesting questions about how these task partition mechanisms evolve. Are the societies applying these mechanisms more efficient than those without task partition? What kind of mechanisms ensure the regulation of workflow and how these systems are able to cope with perturbations and fluctuations? We will investigate most of these with the help of models, but we also made several observations and experiments in nature which we outline here shortly as background information.

5.1.2 The Per capita Paradox

Adaptationist thinking would be ready to tell that sociality has evolved because working in a society is more productive than working alone. However, a fundamental paradox in insect societies is that, as colonies grow, they generally appear to have in fact a lower *per capita* productivity (Michener 1964). Data seems to support the opposite trend of what the armchair thinking would suggest, hence the name "The *Per Capita* Paradox." Possible explanations for this paradox have been numerous, the list includes: bias due to overlooking many small colonies that failed (Michener 1964), kin selection (Hamilton 1964a,b), females of low fitness joining groups, making these groups larger and lowering average fitness (West Eberhard 1978), and a positive effect from protection from enemies (Holldobler and Wilson 1990; Michener and Brothers 1974). Michener (1964) also proposed that, to achieve a small increase in the number of reproductives, a colony must invest much more in terms of workers. This phenomenon may drive the system toward large colony size, especially when workers are long lived. Jeanne and Nordheim (1996), however, challenged Michener's explanation in swarm founding wasps, and they speculated that the quoted phenomenon may be an artifact of the need to lump colonies of different species and development stages to increase sample size. They proposed that per capita output actually increases with swarm size in *Polybia occidentalis*.

Fig. 5.2 Head width against the largest known mature colony size for swarming *Polistinae*. (Reprinted from Karsai and Wenzel 1998 with permission from National Academy of Sciences)

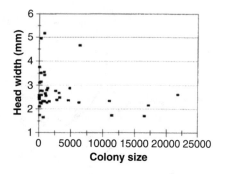

The controversy came from the differences used in data processing and statistics. Instead of calculating per capita variables as Michener did, Jeanne and Nordheim (1996) used an advanced curve-fitting technique including second and third order polynomials that are not biologically interpretable. Recalculating their data and using the idea of Wenzel and Pickering (1991), we showed that larger colonies have a lower variance in productivity (Karsai and Wenzel 1998). We also analyzed this further, using phylogenetic analyses and morphometric data and we concluded that a decrease of variation is a key element in the evolution of social wasps (Karsai and Wenzel 1998). For example, a decrease of head width average and variation with colony size suggests larger colonies operates with a large number of smaller and—with the smaller head—probably less intelligent individuals (Fig. 5.2).

5.1.3 Regulation of Task Partitioning in Metapolybia Wasps

Our previous analyses suggested that in larger colonies a small variation and a steady workflow are the key, and the colonies do not seem to be evolved achieving maximum performance. For understanding the regulation of task partitioning better, we have carried out a set of experiments in Panama's Barro Colorado Island on two species (Karsai and Wenzel 2000). We color coded individual wasps and followed the details of their behavior and we measured the inflow of materials into the nest and also the rate of construction. We also manipulated the material inflow by capturing and excluding foragers temporarily from the workforce or adding and removing material directly to the different components of the system. The collection cycle of a pulp forager was twice as long as construction or water foraging (Fig. 5.3). Both foragers interacted with 5–6 common stomach wasps while they exchanged water and a single pulp load was large enough for 7–8 construction segments carried out by the builders.

These experiments allowed us to reexamine the regulatory schema suggested by Jeanne's (1996) and to parametrize our model. Jeanne suggested a demand driven system (Fig. 5.4a), which was not tested computationally, however. His regulation

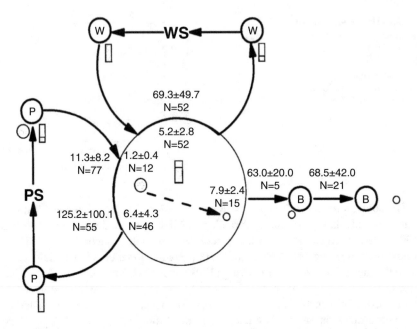

Fig. 5.3 The timing of the construction process and the interaction of workers and materials in nest construction of *Metapolybia* wasps. The big circle in the center represents the pool of colony members. Numbers inside the circle denote the number of wasps involved in the interactions of a single forager. Numbers outside the circle show the duration of different behaviors in seconds. Small circles around a single letter represent a given type of individual: P: pulp forager, W: water forager, B: builder. Thick arrows show the cycle involving pulp source (PS) or water source (WS). Shaded figures represent relative quantities of pulp (circles) and relative water level (box) of a reference individual. Broken arrows within the big circle show pulp sharing. (Reprinted from Karsai and Wenzel 2000 with permission from Springer Nature)

network is a runaway system, which is unable to explain how and why construction starts and ends and why all wasps are not engaged in a construction behavior. Based on our observations, we have constructed a different regulation network (Fig. 5.4b), which we will test in the model section below. The key for this system is provided by the common stomach wasps, which have the flexibility to change into more specialized jobs or just stay generalists and continue to play a role as part of the common stomach. This system suggests high stability, steady material flow, and construction behavior, which are factors we will all test computationally. Our system is also more realistic with assuming the behavioral plasticity of individuals, while in Jeanne's schema the specialists simply switched off and on (which we did not observe), instead of changing to other roles.

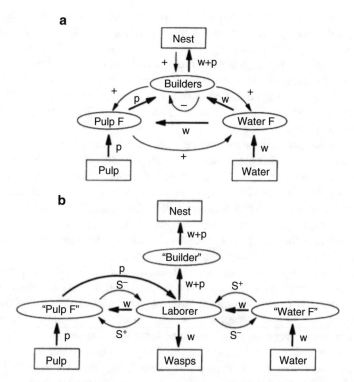

Fig. 5.4 Regulation of the construction behavior in wasps. (**a**) Schema of Jeanne (Figure redrawn from Jeanne 1996, Fig. 11) (**b**) Flow diagram of Karsai and Balázsi (2002). Task groups (oval boxes): Pulp F: pulp forager; Water F: water forager; quotation marks indicate temporal assignment of a given worker, while groups without quotation marks are fixed (predefined). Source or sink of materials is in square boxes; material flow (thick arrow): p: pulp; w: water. "Nest" means that material is built into the nest structure and "Wasps" means the consumption of material by insects. Information flow between task groups (thin arrows): S+: positive or stimulatory effect; S−: negative or inhibitory effect, depending on the water saturation level (S) of the colony. (Reprinted from Karsai and Balázsi 2002 with permission from Elsevier)

5.2   A System Dynamic Model of Task Partition in Social Wasps

5.2.1   Description of the Model

The purpose of the model is to understand how task allocation is regulated in wasp societies. Our goal is to predict general patterns of task allocation, carry out experimental treatments on the model colonies, and compare these predictions to field data. We also predict the global fitness of the modeled colony in terms of its construction efficiency. In addition, we will analyze the sensitivity of our model to key parameters and also carry out perturbation experiments. These operations allow us also to interpret the impact of environmental fluctuations and of sudden changes in the colony structure and in the colony's global fitness. Task regulation will be seen an emergent property of the system without assuming any individual adaptation (such as adapting behavioral thresholds) and without assuming any initial individual differences between the workers. The model is developed in Vensim 5.4 DSS environment and is described in full details in Karsai and Schmickl (2011). The simplified model is deposited online.[1] Below we describe the core of the model in non-technical terms.

The construction of the nest requires pulp and builders; the collection of pulp takes water and pulp foragers. Water foragers provide water for builders and for pulp foragers. Wasps engaged in the studied behaviors can be classified into 4 task groups. Inactive workers (IW): these wasps are generalists and can occasionally change into specialists. Water foragers (WF): specialists, delivering water to the nest. Pulp foragers (PF): specialists, who take water from the colony's common stomach to forage for pulp, and afterwards deliver new pulp to the nest. Nest builders (NB): specialists, who build the pulp into the nest.

Water is not only a building material for the wasps, but it also provides an indirect source of information about the colony's status. This is achieved in the form of a "social crop" or "common stomach" (Karsai and Wenzel 2000), which we believe acts as a global information center in the wasp society. This common stomach is formed because water has to be temporarily stored in the crops of the active members of the wasp society. Water foragers unload water to wasps sitting around the nest entrance and the pulp foragers will beg for water from these wasps (Agrawal and Karsai 2016). This indirect water transfer between these two forager groups is the dominant type of interaction in the nests. The core assumption of our model is that the mean saturation (i.e., the average crop filling) of the common stomach is the key in regulating task allocation (Fig. 5.5):

$$\Omega = \frac{W}{\epsilon N} \tag{5.2.1.1}$$

where W is the amount of water in the colony, ϵ is the maximum crop load of a wasp, N is the colony size, $N = G_I + G_W + G_P + G_B$ (G_I: inactive wasps, G_W: water foragers, G_P: pulp foragers, and G_B: builders).

Water (W) is collected by the water foragers and used by the pulp foragers and builders:

$$\frac{dW}{dt} = \omega \frac{1}{\tau_W} G_W - \phi \frac{1}{\tau_P} G_P - \lambda \frac{1}{\tau_B} G_B \tag{5.2.1.2}$$

where τ-s represent foraging and working durations and λ, ϕ, ω are scaling parameters, respectively. Water foragers are recruited from the inactive wasps and when they abandon their task, they turn into inactive wasps. Recruitment depends on

---

[1] https://sites.google.com/site/springerbook2020/chapter-5.

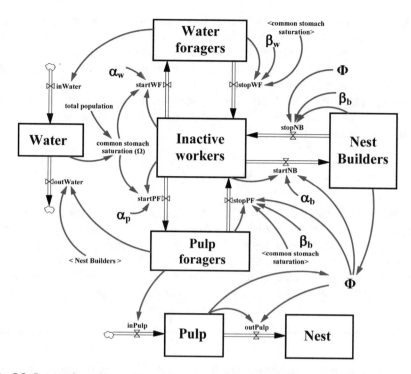

Fig. 5.5 System dynamics model of the task partitioning of the building behavior in wasps. Detailed description of parameters in the text. Rectangular components indicate stocks of quantities, which can accumulate material or task associated workers over time. Double-lined arrows with valves indicate flows of quantities between the stocks, single-lined arrows between the components and variables indicate dependencies (information flows). The small cloud-like symbols indicate sinks and sources (i.e., model boundaries). (Reprinted from Karsai and Schmickl 2011 with permission from Oxford University Press)

the inverse of the common stomach (i.e., a less saturated common stomach increases recruitment), while the abandonment depends on the saturation of the common stomach (i.e., a well-saturated common stomach increases abandonment by water foragers):

$$\frac{dG_W}{dt} = \alpha_W G_I (1 - \Omega) - \frac{\beta_W}{\tau_W} G_W \Omega \qquad (5.2.1.3)$$

where α-s and β-s represent constant recruitment and abandonment rates. The dynamics of the pulp foragers is similar to the water foragers, but is more complex:

$$\frac{dG_P}{dt} = \alpha_P G_I \Omega - \frac{\beta_P}{\tau_P} G_P (1 - \Phi)(1 - \Omega) \qquad (5.2.1.4)$$

Pulp foragers are recruited from inactive wasps and the recruitment is proportional to the saturation of the common stomach (a more saturated common stomach increases recruitment) while the abandonment of pulp forager is inversely proportional to the saturation. The parameters α and β represent the constant recruitment and abandonment rates here as well, but the abandonment also depends on the unloading efficiency (Φ). A pulp forager needs to unload its pulp load to finish the job cycle, and this can only be done if enough builders are present to take that load. This can be described as:

$$\Phi = \min\left[\begin{array}{c} \frac{G_B \eta \tau_\eta}{\tau_b P_P} \\ 1 \end{array}\right] \tag{5.2.1.5}$$

Our field experiments have showed that it takes eight nest builder wasps to take over a single pulp load for the nest building, thus the pulp load taken up by one builder is $\eta = 1$ pulp load/8 wasps. We assume that the act of handing over one pulp load takes a period of $\tau_\eta = 10$ s, thus a fraction of $1/\tau_\eta$ of the pulp is on average available for unloading. The dynamics of the pulp (P_P) depends on this unloading efficiency and the pulp size or amount (σ) as well as the number of the pulp foragers:

$$\frac{dP_P}{dt} = \frac{\sigma}{\tau_P} G_P - \frac{1}{\tau_\eta} P_P \Phi \tag{5.2.1.6}$$

The builder wasps' recruitment rate from the inactive wasps depends on the pulp quantities and is inversely proportional to the unloading efficiency (Φ), while the task abandonment is proportional to Φ:

$$\frac{dG_B}{dt} = \min\left[\begin{array}{c} \frac{1}{\eta \tau_\eta} P_P \\ \alpha_B G_I \end{array}\right] (1 - \Phi) - \beta_B G_B \Phi \tag{5.2.1.7}$$

5.2.2   The Emergence of Task Allocation and the Steady Construction

Simulations of the model system resulted in a life-like dynamics (Fig. 5.6). We started the colonies with an empty common stomach and every wasp originally belonged to the inactive task force. First, a large number of water foragers emerged and these wasps started to fill up the common stomach. As the common stomach became more saturated, pulp foragers emerged next, then the builders. The initial fluctuations turned into a steady workflow. On average, the colony operated with 1.7 water foragers, 3.7 pulp foragers, and 19.3 builders, leaving 9.3 inactive workers. This is a very similar task distribution to what we observed in a similar-size real wasp colony. To test the stability of the model and to better understand the nature of the emergent equilibria, we applied white noise and a strong sinus wave to a single

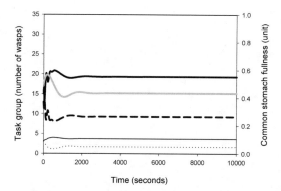

Fig. 5.6 Normal run of the model; both the task groups and the common stomach are relaxed after the initial fluctuations. Dotted thin line: water foragers; solid thin line: pulp foragers; solid thick line: nest builders; broken thick line: inactive wasps; finally, grey solid thick line: common stomach fullness (on the secondary *y* axis). (Reprinted from Karsai and Schmickl 2011 with permission from Oxford University Press)

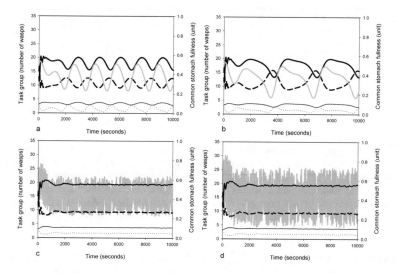

Fig. 5.7 Testing the sensitivity and stability of the model. Upper row: a periodicity implemented using a sine function applied to the recruitment of water foragers. (**a**) Period length of sine function: 500 steps. (**b**) Period length of sine function: 1000 steps. Lower row: uniformly generated random noise applied with ±0.2 (**c**) and ±0.3 (**d**) levels to the common stomach. Dotted thin line: water foragers; solid thin line: pulp foragers; solid thick line: nest builders; broken thick line: inactive wasps; and grey solid thick line: common stomach fullness (on the secondary *y* axis). (Reprinted from Karsai and Schmickl 2011 with permission from Oxford University Press)

parameter perturbation (Fig. 5.7). The high frequency white noise was absorbed by the buffering nature of the common stomach and the sinus wave perturbation was strongly compensated by the rearrangement of the task force.

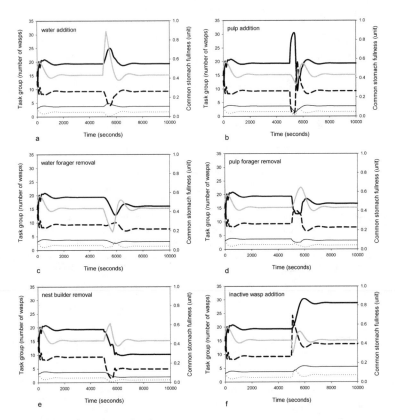

Fig. 5.8 Perturbation experiments, the dynamics of task groups and the fullness of the common stomach. Perturbations were carried out at $t = 5000$; (**a**) water addition, (**b**) pulp addition, (**c**) water forager removal, (**d**) pulp forager removal, (**e**) nest builder removal, (**f**) inactive wasp addition. Dotted thin line: water foragers; solid thin line: pulp foragers; solid thick line: nest builders; broken thick line: inactive wasps; and grey solid thick line: common stomach fullness (on the secondary y axis). (Reprinted from Karsai and Schmickl 2011 with permission from Oxford University Press)

To compare the predictions of the model with experimental data, a series of perturbation experiments were carried out, akin to the published field experiments. The model system was run until it has stabilized, and then one component of the model was suddenly increased or decreased by our simulation settings and then we observed the response of the system. We have made our runs long enough so that we could observe if the system will establish a new equilibrium after the perturbation (Fig. 5.8). The perturbations of the model showed qualitative agreement with the field experiments (Table 5.1). For example, both the addition of water and addition of pulp increased the number of nest builders, via different mechanisms (Fig. 5.8a, b). Addition of water increased the water content and saturation of the common stomach, and this, in turn, has increased the number of pulp foragers, which then transported a larger amount of pulp into the nest, which was processed

Table 5.1 Comparison of the model's predictions with the result of experiments in real colonies. (Reprinted from Karsai and Schmickl 2011 with permission from Oxford University Press)

Manipulation	Studied behavior	Model prediction	Field observation
Add water	Pulp foraging	Increase	2
	Water foraging	Decrease	1, 2
	Nest construction	Increase	2
Add pulp	Pulp foraging	Decrease	1
	Nest construction	Increase	1
Remove WF	Water foraging	Decrease	1
		(Overcompensation of WF later)	1
Remove PF	Pulp foraging	Decrease	1, 2
	Water foraging	Decrease[a]	1, 2
	Nest construction	Decrease	2
Remove builders	Pulp foraging	Decrease	1, 2
	Water foraging	Decrease[b]	2
	Nest construction	Decrease	1, 2

References show agreement with field studies: (1) Jeanne 1996; (2) Karsai and Wenzel 2000. Model prediction is agreement with field studies except
[a]Field data is mixed: non-significant change and or significant decrease
[b]Non-significant change was observed on the field. See Karsai and Schmickl 2011 for more information

by more builders. Based on these results, we proposed that via communicating through an information center (the common stomach) and using a network of worker interactions that establish sets of positive and negative feedbacks, wasps' societies were able to achieve collective information processing and to regulate their colony-level behavior (Karsai and Schmickl 2011).

5.3 The Cost of Task Switching: A Model with Time Delays

5.3.1 Rationalizing the Complication of a New Model

Delayed differential equations (DDE) are commonly used tools to investigate nonlinear dynamics with delay terms (Kuang 1993). The previous model has assumed that the individual wasps are very flexible and that they are able to switch tasks freely. A given forager after abandoning its job will turn into an inactive wasp, where from this state can be recruited into other roles in the next time step. In reality, however, these switches could be more complex and could take more time. For example, a water forager that has collected water for a longer period could become more efficient and has learned where to collect and where to distribute water. After the task change, the wasp needs to adapt to different sets of stimuli. To account for the switching costs, we have combined these sets of efficiency drops into an extra time cost that needs to be paid by the individuals that switch tasks. We assume that during this transition time (reset period) the individual is unproductive. This allows us to investigate the trade-off between adaptation and changes by frequent switches, and the costs of these switches for colony performance.

5.3.2 Description of the Model

Our goal with the new model was to explore the possible effects of task partitioning, noise, and time delays due to transitions between tasks, on colony fitness. To this end, we have developed a simpler basic model compared to the one presented above (Sect. 5.2), but here we have also implemented several new features to investigate the effect of task switching (Fig. 5.9). Here we only describe the core of the model and discuss the main results. Full details can be found in (Hamann et al. 2013).

Fig. 5.9 Overview of different tasks in the system including the transition states of the model. P: pulp forager; W: water forager; L: free laborers; B: builder individual. Light grey arrows show the flow of water, dark gray arrows show the transportation of pulp. The "job change transitions"-box indicates individuals that are in job changing transitions (details on the transitions are in the lower figure). Square boxes with τ-s represent time delays associated with recruitment ($\tau_{LP}, \tau_{LW}, \tau_{LB}$) or abandonment ($\tau_{PL}, \tau_{WL}, \tau_{BL}$). (Reprinted from Hamann et al. (2013) with permission from Springer Nature)

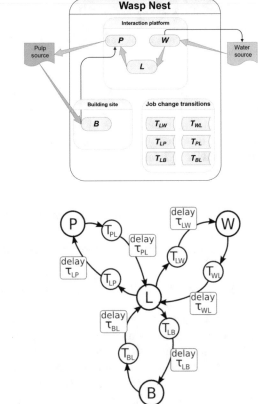

The dynamics of saturation of the common stomach $S(t)$ can be described as water inflow due to water collection by the water foragers ($W(t)$) and water consumption by the pulp foragers ($P(t)$) as well as some random fluctuations:

$$\frac{dS}{dt} = \frac{w_{in}}{L(t)} W(t)(1 - S(t)) - \frac{w_{out}}{L(t)} P(t)S(t) + \xi_{S(t)} S(t) \qquad (5.3.2.1)$$

Here $L(t)$ is the amount of free laborers, w_{in} and w_{out} are the water intake and out-take, respectively, that defines the amount of water contributed by one forager per time as a fraction of one wasp crop (these parameters control the increase and decrease of the saturation of the common stomach). ξ_S is a noise term that models fluctuations in the common stomach and is modeled as a stochastic process (white noise, distributed as $\xi_S \sim nX$ with a standard normally distributed $X \sim N(\mu = 0, \sigma^2 = 1)$ and noise intensity n).

The dynamics of water foragers is described by a recruit term and an abandonment term as well as an additional term that describes the effect of random fluctuations:

$$\frac{dW}{dt} = \Delta T_{LW}(t - \tau_{LW}) - \Delta T_{WL}(t) - \xi_{WL}(t)W(t) \qquad (5.3.2.2)$$

The recruitment process depends inversely on the saturation of the common stomach, plus a stochastic process:

$$\Delta T_{WL}(t) = r_W L(t)\Theta(1 - S(t)) + \xi_{LW}(t)L(t) \qquad (5.3.2.3)$$

Here Θ is a threshold function $\Theta(x) = 1 - 1/(1 + \exp(cx - c/2))$ and r_W as well as c are scaling constants. Abandonment is similarly described, but it is directly proportional to the saturation of the common stomach:

$$\Delta T_{LW}(t) = a_W L(t)\Theta(S(t)) \qquad (5.3.2.4)$$

where a_W is a scaling constant and we omitted the random fluctuation term to avoid duplication of the same process as above.

All this means that, as in the previous model, more water foragers are recruited when the common stomach is in low saturation and the water foragers abandon their task more probably when the saturation is high. The dynamics of pulp foragers is described very similarly, except that their recruitment is directly proportional to the saturation of the common stomach, and they abandon their job more probably when the saturation is low:

$$\frac{dP}{dt} = \Delta T_{LP}(t - \tau_{LP}) - \Delta T_{PL}(t) - \xi_{PL}(t)P(t) \qquad (5.3.2.5)$$

The dynamics of the builders has again a similar simple setup, but their recruitment and abandonment terms are a bit different from those of the foragers:

$$\frac{dB}{dt} = \Delta T_{LB}(t - \tau_{LB}) - \Delta T_{BL}(t) - \xi_{BL}(t)B(t) \tag{5.3.2.6}$$

Here the recruitment term is proportional to the saturation of the common stomach ($S(t)$), the number of pulp foragers ($P(t)$) and free laborers ($L(t)$):

$$\Delta T_{LB}(t) = r_B L(t) P(t) \Theta(S(t)) + \xi_{LB}(t)L(t) \tag{5.3.2.7}$$

The abandonment of builders is described (without a duplication of the random process again) as:

$$\Delta T_{BL}(t) = a_B(B(t) - mP(t))L(t) \tag{5.3.2.8}$$

Builder wasps will abandon their task increasingly with the growing number of free laborers and they stay busy in their task as long as the ratio $B(t)/P(t) \approx m$ is approximately satisfied. The latter is implemented by a summand $-aB(B(t) - mP(t))L(t)$ which goes to zero for $B(t)/P(t) \approx m$. This means that, if there are too many builders ($B(t) > mP(t)$), then the summand is positive. On the other hand, when there are too many pulp foragers ($B(t) < mP(t)$), the summand is negative. A dependence on free laborers L is introduced to diminish the increase of builders in low free laborer situations.

Switching between tasks is always done via the free laborers group and always includes a time delay. During this time delay, the wasps will not work and they also do not contribute to the common stomach. Typically, only a small fraction of wasps is in the transition states, provided the τ-s are small. In general, wasps that were doing a special task X and abandoned that job will switch to transition state TXL. Similarly, wasps that are in the state of free laborer and switch to one of the three special tasks X, first must switch to the transition state TLX. For example, for the transition state from free laborers to water foragers, we get

$$\frac{d\tau_{LW}}{dt} = \Delta T_{LW}(t) - \Delta T_{LW}(t - \tau_{LW}) \tag{5.3.2.9}$$

5.3.3 The Effect of Noise and the Emergence of Periodic Attractors

We treat this system of equations similarly to our first set. We integrate the equations over time, until an equilibrium is reached, then we apply a one-time perturbation in terms of a sudden change of a single component of the system. In case of changes of task groups (swarm fractions), the task groups are not normalized after the perturbation to show the differences more clearly. Then the system equations are integrated until an equilibrium is reached again.

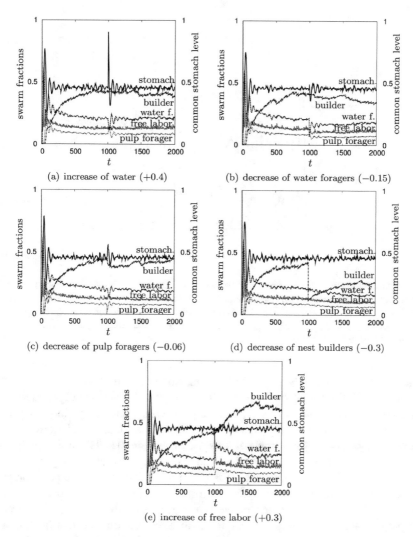

Fig. 5.10 The effect of perturbations on the wasp colony model. A small noise ($n = 0.007$) was applied to all state variables. In case of swarm fraction disturbances, the fractions are not normalized to show relative changes. (Reprinted from Hamann et al. (2013) with permission from Springer Nature)

The system reacts very similarly to the perturbations as our previous model (Fig. 5.10). For example, a sudden increase of water in the common stomach results in a decrease of the water foragers and an increase of pulp foragers and builders. Comparing the standard runs with and without the noise we see that the latter did not change the results dramatically. When the random noise was present, the transitions took longer, but in each case the system had a "before and after" quasi-stable state where the system has fluctuated around. The noise has also decreased

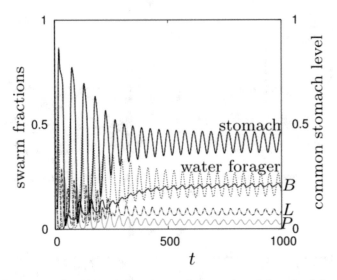

Fig. 5.11 Emergence of stable periodic attractors for strongly unbalanced relations of water extraction to water intake w_{out}/w_{in} combined with high task switch rates Φ. From top to bottom the lines represent common stomach saturation level, fractions of water foragers, builders (B), free laborers (L), and pulp foragers ($P4$), respectively. (Reprinted from Hamann et al. 2013 with permission from Springer Nature)

some fractions, especially that of the builders, but the system reacted similarly to the perturbations. Depending on the parameters, the system does not always converge to an equilibrium (i.e., a set of fixed points) but sometimes shows a periodic behavior. For special parameter settings, the system converges to stable oscillations in some or all state variables. For example, if we set $w_{out}/w_{in} = 10$ and the task switch rates to a high value ($\Phi = 0.3$), this setting results in stable oscillations (Fig. 5.11). The reason that these oscillations emerge is that a small increase of the pulp foragers causes a big increase in water extraction from the common stomach, which in turn elicits an intensive recruitment of water foragers. Due to the increased number of pulp foragers the number of builders will also increase, and this in turn will make sure that the water level of the common stomach starts to drop, which again decreases the number of pulp foragers. Then, due to the low number of water users, and the newly recruited water foragers, the common stomach starts to fill up. This eventually will cause water foragers to revert back to free laborers.

To study the effects of noise and the "water extraction/water intake" ratio, our parameter sweeps were converted into bifurcation diagrams (Fig. 5.12). Lines show the case of fixed points and areas the case of periodic attractors (defined by the minima and the maxima of the oscillations). The diagram starts at the left with values of $w_{out}/w_{in} < 1$, that is, one water forager adds more to the common stomach than one pulp forager will take out, within a given time interval. For about $w_{out}/w_{in} < 6$, we observed single lines and these single lines indicate that the system converges to fixed equilibria. When the ratio is larger ($w_{out}/w_{in} > 6$), we

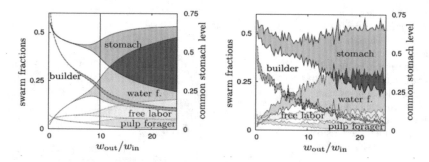

Fig. 5.12 Bifurcation diagram for a noise free (left) and a noisy ($n = 0.014$, right) system. The vertical line at $w_{out}/w_{in} = 10$ corresponds to the setup of the sample run (Fig. 5.11). The task switching rate was high ($\Phi = 0.3$). (Reprinted from Hamann et al. 2013 with permission from Springer Nature)

observed stable periodic behaviors of the system variables. The amplitudes of the oscillations have increased with an increasing imbalance of water extraction to water intake. Applying noise to the system (with a noise intensity $n = 0.014$) concealed this change of the attractor (Fig. 5.12). Only in the case of the water foragers and the common stomach saturation did we observe an increase between the minimum and maximum values with the increasing imbalance of water extraction to water intake. Noise does not add to the variance of the state variables introduced by periodic attractors but rather has a mild damping effect. The emergence of a Hopf bifurcation observed at about $w_{out}/w_{in} = 6$ was to be expected, because periodic attractors in systems described by DDEs and with delayed negative feedback control are quite common.

5.3.4   Effects of Noise and Task Switch Rate on Colony-Level Performance

Colony performance should depend on how frequently the individuals are switching task, because these switches keep the individuals in a jobless transitional state. There is a close relation between the length of time delays τ and the task switch rate ϕ. For moderate task switch rates (depending on the noise level, at about $\phi < 0.1$), increasing the task switch rate reduces the time spent in transient states. Decreasing time delay also reduces the transients. Therefore, varied task switch rates can also be interpreted as changes in the time delays (high task switch rates correspond to short time delays, i.e., low task switch costs). For a simple measure of colony-level performance (γ), we can simply integrate the swarm fraction of the builders over time:

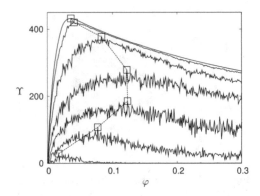

Fig. 5.13 The colony performance as the function of task switching rate and the degree of noise. Noise intensities n from top to bottom: 0, 0.0014, 0.0055, 0.014, 0.028, 0.055, 0.14. Squares mark the maxima of each setting. (Reprinted from Hamann et al. 2013 with permission from Springer Nature)

$$\gamma = \int_{t_1}^{t_2} B(t)\, \mathrm{d}t \qquad (5.3.4.1)$$

This measure will highly correlate with how much material is built into the nest, which is the main goal of the nest construction. At very low switch rates, we have observed an increase in performance at $0 < \phi < 0.05$ (Fig. 5.13). This is due to very long transients. For higher task switch rates (depending on the noise level) the performance decreases again, because higher task switch rates increase the swarm fraction that is "idling" in the transition states. There is an optimum value for apt relations between transient length and task switching intensity. For the noise-free case ($n = 0$) high performance is reached by comparatively low task switch rates. Higher task switch rates could increase the performance, but the transients would be increased as well, and this results in an overall decrease of the performance when $\phi > 0.025$. Noise has an important role here, because increasing noise increases the transients, hence higher task switch rates are advantageous for medium noise intensities. However, optimal task switch rates decrease for the highest noise intensities (Fig. 5.13, lowest 2 lines) because high task switch rates combined with high noise intensities will increase the swarm fraction in transition states.

5.4   Agent-Based Models of Task Partition in Social Wasps: Task Fidelity and Colony Size

5.4.1   Description of the Model

We have also developed agent-based models for the task partition of wasps, in order to demonstrate that the results obtained do not depend on the modeling technique and to investigate several details that could not be modeled efficiently in the top-down models (Karsai and Phillips 2012; Karsai and Runciman 2011a). These models were developed in Java and C++ and we provide a simple demo of

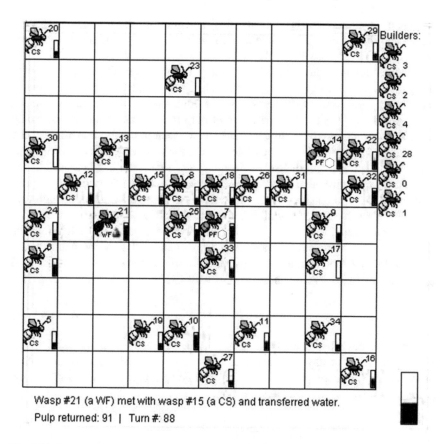

Wasp #21 (a WF) met with wasp #15 (a CS) and transferred water.
Pulp returned: 91 | Turn #: 88

Fig. 5.14 Interactions of workers on the interaction platform (10x10 lattice). Workers: water forager (*W F*) wasps with dark grey color and drop shape; pulp forager (*P F*) wasps with medium grey color and hexagonal shape; Common stomach (*CS*) wasps with grey body and white abdomen. *CS* wasps listed as builders are on the building site constructing the nest and thus not available for interaction. The bar next to a wasp indicates the fullness of its stomach with water. The large bar next to the active platform shows the relative fullness of the *CS*. (Reprinted from Karsai and Runciman 2011a with permission from Taylor and Francis)

the approach.[2] Details of the model are described in the technical papers. Here we only outline the core of the models and focus on the results not investigated in the top-down models discussed before.

This model comprises four hierarchical levels: individuals, interaction platform, building site, and environment (Fig. 5.14). The first two are modeled explicitly while the last two are modeled abstractly (the most important being that the wasps at the building site or at the collection sites are simply taken to spend time away

---

[2]https://sites.google.com/site/springerbook2020/chapter-5.

from the interaction platform). Because this is a bottom-up model, each wasp has a rule set and these rules include interactions with other wasps and the environment (Fig. 5.15).

At the beginning of the simulations all wasps belong to the laborer category, with empty crops, therefore the common stomach is also empty. Soon water foragers emerge and carry water to the nest and the filling up of the common stomach begins, which will trigger the emergence of pulp foragers and builders and finally a steady construction will emerge similarly as we seen in the top-down models. A self-organized task allocation is possible because the wasps are able to change their task group. These changes depend on the number of successful interactions with the environment and with other wasps. For example, if it takes a long time and many interactions to download the foraged material or to prepare for the collection trip, it is highly probable that the wasp is a member of a forager task group which has excess members and it will stop foraging and will become a simple laborer. The laborer wasp can turn into a forager based on its internal state, probabilistically described by a Weibull cumulative distribution function, a function commonly used in problems related to ageing and stress (Weibull 1951):

$$F(x; \alpha, \beta) = 1 - e^{\left(\frac{x}{\beta}\right)^{\alpha}} \qquad (5.4.1.1)$$

This function describes that as the number of unsuccessful interactions increases (x), the probability to stay in the same task decreases. The basic parameters we used for the Weibull function ($\alpha = 5, \beta = 20$) provided a function which emulated the behavior of wasps as we observed in nature.

5.4.2 Predictions of the Model

Results of the agent-based model are very similar to what was found using the top-down models, but we were also able to perform some new investigations in them. The effect of colony size to efficiency and task fidelity are especially important. The general view on insect societies is that larger societies are more efficient and their members are more specialized (probably due to mechanisms related to age dependency or genetical predispositions). Therefore, they rarely switch tasks (high task fidelity, Karsai and Wenzel 1998). Of course, these insights depend on how task fidelity and efficiency are measured. Task fidelity can, in our view, be monitored by counting how many foraging cycles an individual carries out before changing to another task. When a water forager completes a foraging cycle (for example, to fly out for water, collect water, fly back to the nest, and give away the collected water), this forager needs to decide whether to start a new foraging fly again (thereby staying in the same role) or not (i.e., changing into a laborer). The length of the consecutive "stay as water forager" decisions (L_w) and on the number of occasions being a forager (i) are measured to obtain the fidelity of the water foragers (F_w):

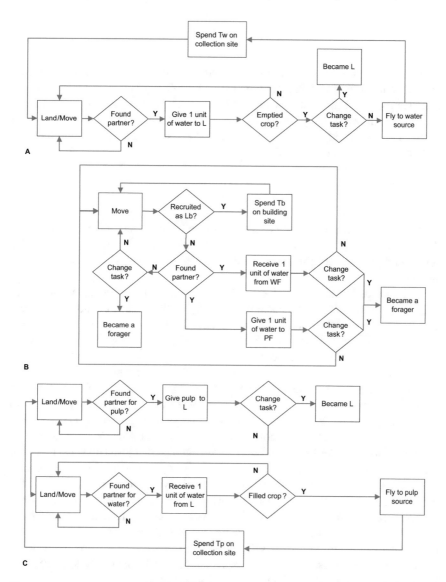

Fig. 5.15 Process overview of the model. Water foragers (WF) (a) and pulp foragers (PF) (c) can change to laborers (L) (b). Laborers can accept pulp and turn into builders (L_b), receive and give water (form the common stomach), and can change to foragers. Each task group has a working cycle that includes landing/moving decisions and actions and the time needed to carry out the given jobs: T_w: time needed for collecting water; T_p: time needed to collect pulp; T_b time needed to carry out construction. (Reprinted from Karsai and Phillips 2012 with permission from Elsevier)

$$F_w = \frac{\sum_0^i L_w}{i} \tag{5.4.2.1}$$

The fidelity of pulp foragers is calculated the same way and, during each run, the maximum length of being continuously a pulp forager (L_p) or a water forager (L_w) is recorded. If this value is large, this indicates that there are more specialized foragers (high task fidelity) in the colony. If the value is small, this indicates that foragers are frequently reverting to laborers. We also used a very simple measure for the construction efficiency E:

$$E = \frac{P_u - P_d}{N_a / T} \tag{5.4.2.2}$$

where P_u is the amount of pulp that has been delivered to the interaction platform, P_d is the amount of pulp discarded because of a lack of free laborers on the nest; N_a is the number of active wasps, and T is the number of time steps since the simulation started.

With increasing colony size, the number of foragers has somewhat increased, but only by a small amount and not proportionally to the growth of the colony size. The fluctuations of the common stomach have decreased, but most of the added wasps did not increase the average value of the common stomach or the number of foragers (Fig. 5.16a). The efficiency of the colony has significantly increased with the colony size, but this increase was really more pronounced at the lower numbers. Because the increase in the pulp foragers were much smaller than the increase of the colony size, such moderate increase in the efficiency is easy to explain (Fig. 5.16b). More interesting was our finding when the colony size was only 10, i.e., very low. In this case more than a quarter of the colonies have failed (for example, each member became a water forager and so there was no wasp to accept the water, thus the system got stuck and failed). It first seems that this is a modeling error, but in fact such errors do support the validity of our model. In nature, the wasps that use the common stomach mechanism are multiplying via colony fission (swarming). Many individuals leave the old nest and colony to start a new nest and new colony, therefore they practically never go under the critical colony size, where the mechanism would fail them. Other species of wasps start the colony cycle by a single queen and these species did not evolve the mechanism of task partitioning based on the common stomach regulation (Karsai 1999).

We were also able to show that task fidelity (i.e., specialization) significantly increases with colony size (Fig. 5.16c). We observed this pattern in the model without implementing those mechanisms that were supposed to ensure this pattern, such as genetical dispositions, learning, age related mechanisms, and so on (Karsai and Wenzel 1998). Our model included none of these mechanisms. Each wasp was equal and started out the same way and there was not any learning or physiological change possible. So why the observed phenomenon? The larger task fidelity emerged automatically in the larger colonies, because the material flow showed less fluctuations and the wasps had a larger probability to meet with a

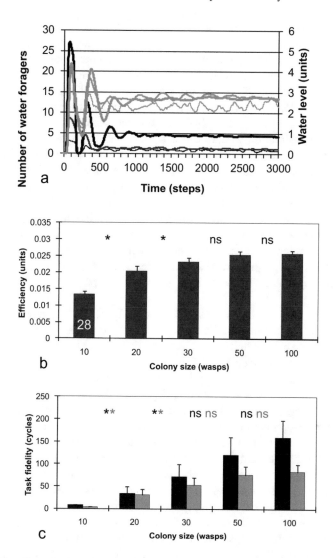

Fig. 5.16 The effect of colony size on colony-level performance: (**a**) the number of water foragers (black lines) and the water level of the common stomach (grey lines). Thicker lines mean higher value of the parameters ($N_a = 10, 30$ and 100). (**b**) colony-level efficiency. Numbers inside the columns indicate the number of failed colonies. (**c**) task fidelity of water foragers (black columns) and pulp foragers (grey columns). Average values and standard deviations were calculated from 100 parallel runs (failed colonies not considered). Significance level: $^*p < 0.05$ (color correspond to columns); *ns*: non-significant, Mann–Whitney U test. (Reprinted from Karsai and Phillips 2012, with permission from Elsevier)

partner who accepted their interactions, and therefore make their work a success. No special mechanism needs to evolve to ensure the emergence of specialists, the flexible individuals carrying out the same behavior become more specialized in an environment where more individuals exist (Karsai and Phillips 2012).

As presented before, the colonies operate with a small number of foragers and increasing the colony size will not increase the number of foragers proportionally. This contradicts the simple logic that more foragers would provide more materials and then the construction would be faster. Sure, but this is not happening in real colonies either. To test how efficient the various colonies with different task groups can be, we carried out extensive parameter sweeps on a simplified model where the workforce combination was fixed (so the wasps could not change task) and the efficiency was measured. These runs focused on small ($N = 20$) and medium ($N = 40$) sized colonies and all possible worker combinations were simulated 20 times each (the average of these 20 colonies is represented by one point in Fig. 5.17. The efficiency might also depend on how much time is needed for collecting the materials. To simulate this, we also considered two scenarios where the pulp and water collecting needed the same time and when the pulp collecting needed four times more.

From these runs we can see several trends (Fig. 5.17). The most important is that there is an optimal mix of the workforce (i.e., the graphs have a peak). Larger colonies could be more efficient, because, e.g., duplicating the colony size can result in more construction. Independent from the colony size, about half of the colony should be a common stomach wasp to get a colony with maximum efficiency. The other half of the colony will forage and depending the time required for foraging, the types of foragers are adjusted accordingly. For example, if pulp foraging takes longer, then the optimal composition contains more pulp foragers and less water foragers. We can conclude that the common stomach mechanism works best if the colonies are larger and these colonies operate with a smaller number of more specialized foragers and a large number of common stomach wasps.

Why has this mechanism evolved? We believe that it is the result of several trade-offs. Leaving the nest for foraging is inherently very dangerous. The wasp then becomes a solitary animal with no defense against birds and other predators. At the nest, however, the wasps have a common defense against predators. Frequent traffic would also call the attention of potential predators for the nest. The colony probably will thus lose less individuals if it uses a small number of specialized foragers. These specialized foragers can make foraging faster, because they become more experienced. Keeping large number of wasps in the nest as common stomach wasps has not only the benefit of increasing the interactive partners for the foragers, but these wasps have important secondary functions, to guard the nest *en masse* and to provide a source pool for foragers when some of them get killed (Karsai and Runciman 2009, 2011a,b).

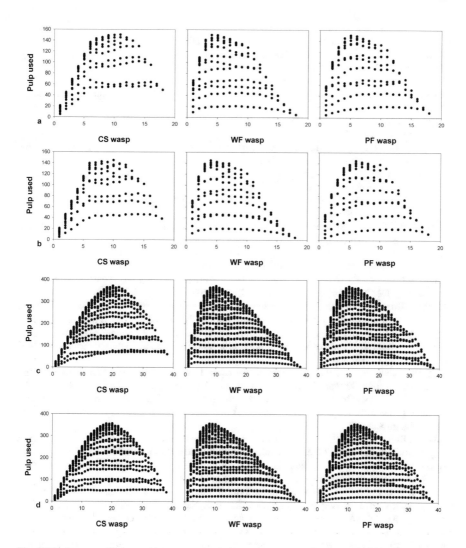

Fig. 5.17 Average efficiency (dots represent the average amount of pulp used in 20 parallel runs) of different workforce combinations in small ($N = 20$) (**a**, **b**) and larger ($N = 40$) (**c**, **d**) colonies. Each panel row describes the efficiency distribution of the tree tasks. Scenarios with different foraging times: (**a**, **c**): $T_p = T_w = 4$; (**b**, **d**): $T_w = 2, T_p = 8$. (Reprinted from Karsai and Runciman 2011a with permission from Taylor and Francis)

References

Agrawal D, Karsai I (2016) The mechanisms of water exchange: the regulatory roles of multiple interactions in social wasps. PLoS One 11(1):e0145560. https://doi.org/10.1371/journal.pone.0145560

Barlow R, Proschan F (1975) Statistical theory of reliability and life testing. Holt, Rinehart and Winston, New York

Hamann H, Karsai I, Schmickl T (2013) Time delay implies cost on task switching: a model to investigate the efficiency of task partitioning. Bull Math Biol 75:1181–1206

Hamilton WD (1964a) The genetical evolution of social behaviour. J Theoret Biol 7:1–16

Hamilton WD (1964b) The genetical evolution of social behaviour II. J Theoret Biol 7:17–52

Holldobler B, Wilson EO (1990) The ants. Belknap, Cambridge

Jeanne RL (1996) Regulation of nest construction behaviour in Polybia occidentalis. Anim Behav 52:473–488

Jeanne RL, Nordheim EV (1996) Productivity in a social wasp: per capita output increases with swarm size. Behav Ecol 7:43–48

Karsai I (1999) Decentralized control of construction behavior in paper wasps: an overview of the stigmergy approach. Artif Life 5:117–136

Karsai I, Balázsi G (2002) Organization of work via a natural substance: regulation of nest construction in social wasps. J Theoret Biol 218:549–565

Karsai I, Pénzes Z (1993) Comb building in social wasps: self-organization and stigmergic script. J Theor Biol 161:505–525

Karsai I, Pénzes Z (1998) Nest shapes in paper wasps: can the variability of forms be deduced from the same construction algorithm? Proc R Soc Lond B 256:1261–1268

Karsai I, Pénzes Z, Wenzel JW (1996) Dynamics of colony development in Polistes dominulus: a modeling approach. Behav Ecol Sociobiol 39:97–105

Karsai I, Phillips MD (2012) Regulation of task differentiation in wasp societies: a bottom-up model of the "common stomach". J Theor Biol 294:98–113

Karsai I, Runciman A (2009) The effectiveness of the "common stomach" in the regulation of behavior of the swarm. In: Troch I, Breitenecker F (eds) Proceedings MATHMOD 09 Vienna full papers CD volume, 6th Vienna conference on mathematical modelling. February 11–13 2009, Vienna University of Technology, Austria. ARGESIM report no 34:851–857. ARGESIM Publishing House, Vienna

Karsai I, Runciman A (2011a) The common stomach as a center of information sharing for nest construction. In: Kampis G, Szathmary G, Karsai I, Jordan F (eds) Advances in artificial life. Darwin meets von Neumann. 10th European conference, ECAL 2009 (Lecture notes in artificial intelligence, vol 5777 subseries: lecture notes in computer science, vol 5778). Part II. Springer, Berlin, pp 350–357

Karsai I, Runciman A (2011b) The "common stomach" as information source for the regulation of construction behavior of the swarm. Math Comput Model Dyn Syst 18:13–24

Karsai I, Schmickl T (2011) Regulation of task partitioning by a "common stomach": a model of nest construction in social wasps. Behav Ecol 22:819–830

Karsai I, Wenzel JW (1998) Productivity, individual-level and colony-level flexibility, and organization of work as consequences of colony size. Proc Natl Acad Sci USA 95:8665–8669

Karsai I, Wenzel JW (2000) Organization and regulation of nest construction behaviour in Metapolybia wasps. J Ins Behav 13:111–140

Kuang Y (1993) Delay differential equations: with applications in population dynamics. Academic Press, Boston

Michener CD (1964) Reproductive efficiency in relation to colony size in hymenopterous societies. Insect Soc 11:317–341

Michener CD, Brothers DJ (1974) Were workers of eusocial hymenoptera initially altruistic or oppressed? Proc Natl Acad Sci USA 71:671–674

Schmickl T, Karsai I (2018) Integral feedback control is at the core of task allocation and resilience of insect societies. PNAS 115(52):13180–13185. https://doi.org/10.1073/pnas.1807684115

Weibull W (1951) A statistical distribution function of wide applicability. J Appl Mech 9:293–297

Wenzel JW, Pickering J (1991) Cooperative foraging, productivity, and the central limit theorem. Proc Natl Acad Sci USA 88:36–38

West Eberhard MJ (1978) Polygyny and the Evolution of Social Behavior in Wasps. J Kansas Entomol Soc 51:832–856

Chapter 6
Ants and Bees: Common Stomach Regulation Provide Stability for Societies

Abstract The common stomach regulation system, albeit originally found in wasp societies, could be applied to other insect societies as well. We generalized the concept of the common stomach and applied this new concept to describe task allocation mechanisms for *Ectatomma* ants and honeybees. In these systems, the in- and outflow of the essential materials are regulated by the workforce, and the workforce itself is regulated by the saturation of the materials in the colony. The stability and resilience of colonies are ensured by these feedback loops at various scales. Our sensitivity tests and a comparison of the model predictions to experimental data shows that this regulation appears to be a key to the universal success of insect societies. We show that this mechanism is very robust in general and works well in different environments. We also tested several evolutionary scenarios concerning plant-pollination vs. honeybee colony overwintering, and found that pollen hoarding would be beneficial for the plant and for the short time success of the colony, but it would decrease the honey storages at the same time, which essential for overwintering. A regulation based on the common stomach gives the highest fitness overall.

6.1 Background

In the previous chapter we have demonstrated how a common stomach-based regulation system allocates workforce to ensure steady construction work in wasps. The core in this regulation system is one kind of essential material (water in wasps), which is temporarily hold in the crops of some of the workers and, due to the frequent exchanges of materials among the colony members, the individual crops form a connected system, a "common stomach." The degree of fullness of this common stomach governs the task force distribution, which in turn regulates the water inflow and outflow, and hence the fullness of the common stomach in the colony. This regulation system seems to be very specific about the life of the group of the given wasp species. Can we find a similar regulation system elsewhere? Can we generalize the common stomach approach? Can we describe detailed interactions that ensure a common stomach property?

© Springer Nature Switzerland AG 2020
I. Karsai et al., *Resilience and Stability of Ecological and Social Systems*,
https://doi.org/10.1007/978-3-030-54560-4_6

Efficient allocation of workers to appropriate tasks that ensure a near-optimal performance is a crucial challenge for all insect societies. Colonies of wasps, ants, termites, and honeybees commonly operate with a high number of workers and can even have a still higher number of brood in their nests (Wilson 1971; Holldobler and Wilson 2008). To support the operation of these vast and dense colonies and to adapt to the always-present environmental fluctuations, social insect colonies need operating mechanisms that allow for a flexible regulation and an efficient supply of nutrients, oxygen and waste products. Various mechanisms have been proposed for task regulation in social insects, for example, those reviewed in (Robinson 1992; Anderson et al. 2001; Beshers and Fewell 2001). Most of the mechanisms suggested were based on processes that basically use the enhancing (by a so-called snowball effect) of the positive feedback that boosts insignificant starting differences from a rather evenly distributed, closely similar set of agents, into distinct sets of task groups or into spatially located groups. Essentially, this positive feedback will amplify the initial random noise that exists in the starting conditions or is applied at runtime. This was shown to operate in the well-accepted mechanisms of ant pheromone trails (Pratt et al. 2002; Beckers et al. 1990) or food source selection (Seeley et al. 1991) and in the aggregation behavior of several organisms (Dussutour et al. 2010). The nonlinear responses we can observe in these systems are usually attributed to the strong effects of this *positive* feedback.

The mechanism we found in wasps is in many ways the opposite. The common stomach system is based on the prominence of strong negative feedbacks in regulating task selection. Negative feedback is commonly associated with a steady state, where the system can maintain its status quo instead of changing it. True, resilience, and maintaining a stable material flow are also key elements of these societies (Middleton and Latty 2016), but are much less studied than the mechanisms of the divergence of these societies (such as the division of labor). Resilience of the system also means that the system has to be very reactive and be able to cope with changing demands and perturbations. The resilience and the reactivity, in the short term, need to be also compatible with other, long-term mechanisms such as the age-polyethism regime (see Sect. 6.3.1) found in honeybees (Huang and Robinson 1996).

In the following sections we will show two examples where a generalized common stomach regulation is able to explain a colony performance previously explained by different mechanisms. We want to stress that resilience and stability ensured by the regulatory role of a strong negative feedback are a key element of the decentralized control of the division of labor (Middleton and Latty 2016). This mechanism ensures that the system is buffered against small changes (Karsai and Schmickl 2011; Karsai and Phillips 2012; Agrawal and Karsai 2016) and will pay the cost of task switching when this is really needed (Hamann et al. 2013).

6.2   Task Allocation in the Colony of *Ectatomma ruidum* Ants

6.2.1   The Regulatory Role of Density of Prey and Dead Corpses

Ectatomma ruidum is a common ant in Central and Northern South America's forests and plantations. It preys on a wide variety of arthropods (Lachaud et al. 1990; Schatz et al. 1995). In general, the foragers of this ant species hunt solitarily. The ant attacks an insect, kills it with her stinger, and takes the corpse into the nest (Lachaud et al. 1984; Lachaud 1985). However, under some conditions, this solitary hunting is replaced with a more complex collective strategy (Fig. 6.1). Schatz et al. (1997, 1999a,b) found that the hunting activity depends upon prey weight and size and also upon the available quantity of prey in the environment. In case of a high density of the available prey, the hunting task is partitioned into two serialized tasks: Some ants act as "stingers," which are specialized in killing the prey and other ants act as "transporters," which are specialized to collect and transport the prey corpses to the nest. A similar pattern was also found and described in *Pachycondyla caffraria* (Agbogba and Howse 1992). The specific behavior of *Ectatomma ruidum* is described in detail by Lachaud (1990): The stinger grabs a prey animal with its mandibles and sting it until the prey becomes motionless, then the stinger drops the corpse to the ground and seeks for another prey. In some instances, the transporters can solicit a corpse from a stinger if the stinger still holds its victim. However, it is more common that transporters pick up dead corpses from the ground and transfer

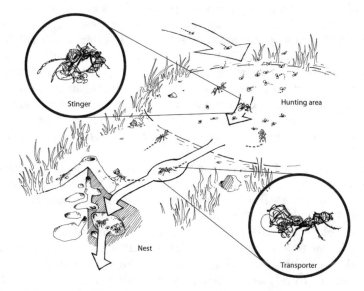

Fig. 6.1 Artistic representation of the collective foraging system established by a colony of *Ectatomma ruidum* by P. Klash. (Reprinted from Schmickl and Karsai 2014 with permission from PLOS)

these corpses to the nest. Usually, the transporter immediately returns to the hunting site for picking up another corpse after a successful delivery of a copse to the nest.

Schatz et al. (1999a,b) and Theraulaz et al. (2002) have constructed two mathematical models to describe the task partitioning of *Ectatomma ruidum*. These models are based on the idea of a simple response threshold model (Plowright and Plowright 1988; Theraulaz et al. 1998). According to this model, if the intensity of a stimulus associated with a specific task exceeds the response threshold of a given worker, then the worker will engage in that task. In such models, task partitioning emerges by the dynamic interactions of the interacting subtasks: for example, performing subtask A can decrease the stimulus intensity of subtask A, but it can also increase the stimulus intensity of subtask B. This way, through well-designed threshold-curves and task interactions, self-organizing patterns of division of labor can emerge and provide predictions similar to the experimental data. We proposed an alternative model of these phenomena, modeled by Schatz et al. (1999a,b) and (Theraulaz et al. 2002) where instead of using nonlinear stimulus-response threshold functions, we rely on the regulatory work of the common stomach mechanism (Schmickl and Karsai 2014).

6.2.2   A Common Stomach Model of the Ants

To implement the common stomach regulation of different food materials, we needed to redefine or generalize what the "common stomach" originally means. The common stomach in wasps was considered as a social crop, where water is temporarily stored in the stomachs of wasps. In more general terms, the common stomach is a well-defined area (space) where a substance (material) can accumulate and the spatial density (or the "degree of fullness of the common stomach") of this material can be estimated locally by the individual workers through simple cues. This local estimate then affects the workers' behavior, which in turn, changes the accumulation dynamics (accumulation rate) of the material in the common stomach. This way, feedback loops can emerge and these provide a robust regulation of material flows and task partitioning. In case of the ants, this approach allows us to redefine the density of the prey in the hunting area as a "saturation of the common stomach."

This generalization allowed us to model the task partitioning of the *Ectatomma ruidum* ant using a common stomach regulation system. We only show the core of the model here. The detailed description of the model is in (Schmickl and Karsai 2014) and the model is available online.[1]

The ant colony was modeled by systems dynamics to ensure the conservation of mass. A resulting Stock and Flow representation of the model provides a visualization of information and material flows (Fig. 6.2). Both the Stinger (S) and Transporter (T) ants are recruited from the Undecided hunters (U), and these

---

[1] https://sites.google.com/site/springerbook2020/chapter-6.

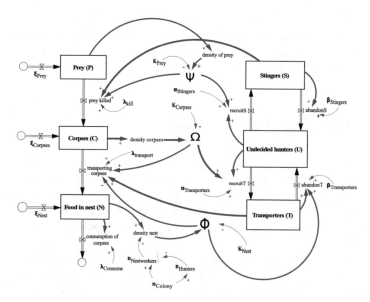

Fig. 6.2 Stock and flow representation of the model of task partitioning of *E. ruidum*. Boxes (stocks) represent quantities of ants or food; double arrows represent material flows and single arrows represent information flows. Capital Greek symbols indicate common stomachs. Reprinted from Schmickl and Karsai (2014) with permission from PLOS

specialists are then reverted back to undecided when they abandoned their job. The recruitments and abandonments of jobs are regulated by the fullness of three common stomachs: Ψ: saturation of prey in the hunting area; Ω: saturation of corpses (killed prey) in the same area; Φ: saturation of food in the nest. The saturation of these common stomachs depends on the size of the task groups, which process these materials and on some external factors, such as the emergence of prey in the hunting area. Albeit the ant system is more complicated than the wasp one, a similar set of equations can be used to describe the task allocation. For example, the dynamics of stinger ants (S) is this:

$$\frac{dS}{dt} = \alpha_{Stingers} \Psi(t) - \beta_{Stingers} S(t) \qquad (6.2.2.1)$$

The recruitment term (first term) depends on a recruitment scaling constant (α), the number of undecided hunters ($U(t)$), and the saturation of the prey on the hunting area ($\Psi(t)$). The abandonment term (second term) depends on a recruitment scaling constant (β) and the number of the stingers ($S(t)$). The dynamics of the transporter ants are similar, but depend on two of the common stomachs:

$$\frac{dT}{dt} = \alpha_{Transporters} \Omega(t) U(t) - \beta_{Transporters} \Phi(t) T(t) \qquad (6.2.2.2)$$

The recruitment term depends on the recruitment constant (α), the number of undecided hunters ($U(t)$) as well as the common stomach representing the saturation of dead corpses in the hunting area ($\Omega(t)$). The abandonment depends, besides the scaling constant (β) and the number of transporters ($T(t)$), on the saturation of the nest with food (i.e., the common stomach $\Phi(t)$). Undecided ants (U) are simply the remaining population of the total number of ants that participate in food gathering (G): $U(t) = G - S(t) - T(t)$.

6.2.3   Predictions of the Ant Model

The model provides predictions that agrees with the experimental data and results of the previous models (Schatz et al. 1999a,b; Theraulaz et al. 2002). When prey influx to the hunting area was constant and high, many undecided hunters have specialized into stingers and transporters and the system converged to an equilibrium if the food was not accumulating in the nest (no more corpses arrived than were consumed, Fig. 6.3). In an experiment the cited authors suddenly put 80 new prey items in the hunting area and then closed the further influx of the prey. This resulted in a sudden emergence of the two specialized tasks, which after a peak started to decay back into the undecided hunters when the prey was running out. Our model was also able to simulate this colony-level response (Fig. 6.4). This experiment shows that the ant colony is reacting very well to the changes of prey availability. When the prey is scarce, the "undecided" generalist hunters search for the prey and should they find one, they kill it and bring it back to the nest. However, when many prey individuals are present, they might be there only for a very short time. Killing many of them will impede their escape, therefore the emergence of specific killers could increase the food input into the colony. This could be an adaptive response even if there is a possible cost of this behavior, because the transporter might in fact not find all the dead corpses lying around on the hunting ground. But specializing for transportation also saves time for the ant, because these ants do not need to chase and kill preys, just find dead corpses and carry them to the nest.

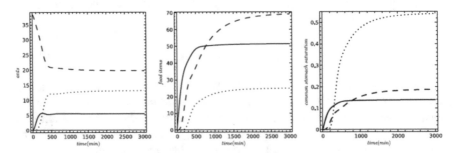

Fig. 6.3 Dynamics in case of a steady prey influx to the hunting area. Equilibria emerge in worker group sizes (left), food stocks (middle) and common stomach saturation levels (right). Left figure: Solid line: S, dashed line: U; dotted line: T. Middle figure: Solid line: P, dashed line: C; dotted line: N. Right figure: Solid line: Ψ, dashed line: Ω; dotted line: Φ. Reprinted from Schmickl and Karsai (2014) with permission from PLOS

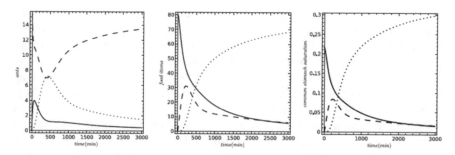

Fig. 6.4 Predicted colony reactions to a small number of prey ($P(0) = 80$) introduced at the start of the experiment, without further prey influx, leads to a pulsed response of the colony. Left figure: Solid line: S, dashed line: U; dotted line: T. Middle figure: Solid line: P, dashed line: C; dotted line: N. RSolid line: Ψ, dashed line: Ω; dotted line: Φ. Reprinted from Schmickl and Karsai (2014) with permission from PLOS

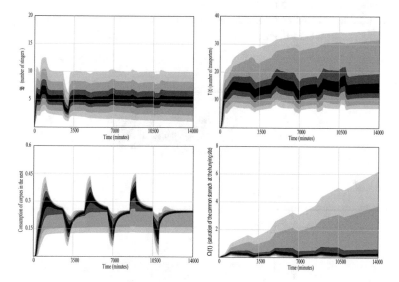

Fig. 6.5 Sensitivity analysis of the common stomach model of the collective foraging of *Ectatomma ruidum*. All key parameters of the model were varied in a random uniform manner within the range of $\pm 50\%$ around their default values and combined using a Latin Hypercube sampling method. The extra perturbations: Prey influx increased (500–21,000 min); prey influx decreased (2500–23,000 min); influx of corpses increased (4500–25,000 min); corpse influx decreased (6500–27,000 min); influx of corpses in the nest increased (8500–29,000 min) and influx of corpses in the nest decreased (10,500–211,000 min). The black region contains predictions of 33% of all 1000 simulation runs. The dark gray region contains 66%, the medium gray region contains 95%, and the light gray regions contains all predictions. Reprinted from Schmickl and Karsai (2014) with permission from PLOS

If both models, the one based on stimulus-response curves (Schatz et al. 1999a,b; Theraulaz et al. 2002) and the second one based on the common stomach regulation (Schmickl and Karsai 2014), are able to predict the same experimental data, then how is it possible to see any differences between the two approaches? How can one decide? To investigate the structural integrity and stability of both approaches, we carried out a series of sensitivity analyses. These analyses combine and change the parameter values of the model to test its sensitivity to initial conditions and for structural integrity (whether the model predicts unfeasible results, such as, e.g., a negative number of ants). We used Vensim's built-in sensitivity analysis tools (Eberlein and Peterson 1992) and performed 1000 simulation runs. The initial values of parameters were sampled and combined randomly (random uniform distribution) with the Latin Hypercube sampling method. Moreover, we also included an external perturbation experiment we carried out in the model (Schmickl and Karsai 2014). Our system did not have unfeasible predictions, even in the case of strong perturbations, and it compensated well for the effect of these perturbations (Fig. 6.5). Carrying out a sensitivity analysis without any perturbation on the model, based on the "response threshold model" (Theraulaz et al. 2002) showed, however, that many initial parameter combinations would lead to non-feasible results (Fig. 6.6). From this, we concluded that our model, the one based on common stomach regulation, has more structural stability than the other, the response threshold model.

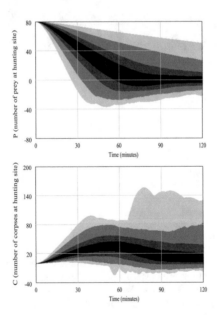

Fig. 6.6 Sensitivity analysis of the original model of (Theraulaz et al. 2002) about the collective foraging of *Ectatomma ruidum*. Using Vensim's built-in sensitivity analysis tool, we have varied the model's key parameters (recruitment rates, abandonment rates, stinging success rate, and transportation rates) in a uniform random manner within the range of ±50%SS around the default values, using a Latin Hypercube sampling method. The black region contains predictions of 33% of all 1000 simulation runs. The dark gray region contains 66%, the medium gray region contains 95%, and the light gray regions contains all predictions. Reprinted from Schmickl and Karsai (2014) with permission from PLOS

6.3 Task Allocation of Foraging in Honeybees

6.3.1 Regulation Mechanisms in Honeybees

The best-known mechanism for the division of labor in honeybee colonies (*Apis mellifera L.*) is called "age polyethism" (Lindauer 1952; Seeley 1982). Under this concept, the workers determine their preferred engagements in specific tasks based on their age. This largely happens because of the physiological changes of the individuals. As they age, this makes them more suited for given tasks. For example, wax producing organs only function in younger bees for a few days, therefore these bees participate in building wax combs. And so on. The resulting worker allocation mechanism reacts slowly to the changing environment and colony needs, because generally the age distribution does not change rapidly. The age cohorts tend to progress very linearly: old forager bees will die, to be replaced by mature bees and new bees emerge to replace the now matured cohorts. This slow age-dependent task allocation mechanism is augmented by other mechanisms that allow for faster reactions to environmental changes in which workers can self-regulate the division of labor (Holldobler and Wilson 2008). Besides the well-known dance-based regulation mechanism (Frisch 1927; Seeley 1992), nurses were found to assess chemical stimuli about the hunger state of the brood which determines their engagement with the feeding task (Huang and Otis 1991a,b). In an ultimate reasoning, it can be assumed that natural selection has favored those flexible self-regulatory proximate mechanisms that are scalable and robust enough to promote the survival and reproduction of colonies.

Honeybees have developed several regulation mechanisms to ensure an efficient foraging and allocation of nutrients that can work in fast-changing conditions. For example, when pollen shortage occurs (caused, e.g., by several days of rain), honeybee colonies tend to perform several compensating strategies in parallel. Pollen foraging becomes significantly enhanced at the expense of nectar foraging (Lindauer 1952), the number of brood and the feeding of the brood are reduced significantly (Blaschon et al. 1999; Schmickl and Crailsheim 2002; Schmickl et al. 2003). Cannibalism of eggs and larvae enables reallocation of the proteins already spent on brood care in order to rear a few well-fed larvae while keeping the food supply of the queen on an almost consistent level (Schmickl and Crailsheim 2001; Schmickl et al. 2003). All these compensating strategies require that individual workers change their behavioral patterns and adapt their task selection in the short term (Riessberger and Crailsheim 1997).

Our main hypothesis is that a network of feedback loops regulates the short-term adjustments of the task allocation of foraging and nursing in honeybees. We show that these regulatory loops can be described by a common stomach system similarly to what we found in wasps and ants, and that this regulation system does not only affect work allocation and nutrient stores in the colony, but also the brood age composition through cannibalism, which ultimately affects the long-term age polytheism regulation system working more slowly.

6.3.2 The Common Stomach Model for Honeybees

Honeybee society is more complex than are either the ant or wasp societies we have modeled in the previous chapters. Therefore, the model of honeybee foraging will be quite complicated. Here we only show the core of this model, and all details are described in (Schmickl and Karsai 2016, 2017). The model is available online.[2]

At the core of the model, there are two common stomachs, generalized (as in the ant case) as saturable storage systems for a specific substance. The first common stomach concerns the saturation of workers and brood with proteins (Ω), which affects many elements of the regulation network including task allocation, protein consumption, brood development, and so on. The second common stomach concerns the level of saturation of nectar stores (Φ), which affects the ratio of loaded to unloaded storer bees, which in turn regulates the recruitment of nectar foraging bees (Fig. 6.7). These two specialized foragers are drawn from the same limited pool of inactive potential forager bees and the regulation of this workforce ensures the adaptability of the colony to ever-changing food availability. The dynamics of the foragers can be described in a similar fashion as we did for the ant and wasp colonies.

The change of pollen foragers F_{pollen} can be modeled as this:

$$\frac{dF_{pollen}}{dt} = \alpha_{pollen} F_{inactive}(t)(1 - \Omega(t)) - \beta_{pollen} F_{pollen}(t)\Omega(t) \qquad (6.3.2.1)$$

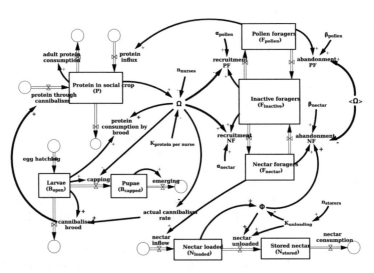

Fig. 6.7 Stock and flow model of the task allocation of foraging in honeybees. Boxes (stocks) represent quantities of individuals (adult and larva bees) or food; double arrows represent material flows and single arrows represent information flows. Capital Greek symbols indicate common stomachs. Reprinted from Schmickl and Karsai (2017) with permission from PLOS

---

[2]https://sites.google.com/site/springerbook2020/chapter-6.

where α and β are scaling constants of recruitment (first term) and abandonment (second term), respectively. Recruitment is drawn from the inactive forager group ($F_{inactive}$) and is proportional to the emptiness of the common stomach $(1 - \Omega(t))$. Abandonment is proportional to the fullness of the common stomach $(\Omega(t))$ and pollen foragers abandoning their job will revert to inactive foragers. The dynamics of the nectar foragers (F_{nectar}) can be described similarly, but depends on both common stomachs:

$$\frac{dF_{nectar}}{dt} = \alpha_{nectar} F_{inactive}(t)\Omega(t) - \beta_{nectar} F_{nectar}(t)\frac{1}{2}(1 - \Omega(t) + \Phi(t)))$$

$$(6.3.2.2)$$

The process of recruitment for nectar foraging is directly proportional to the protein saturation of the colony $(\Omega(t))$, while the process of abandonment of nectar foraging is directly proportional to the average of the two common stomachs $(1 - \Omega(t)) + (\Phi(t))$. Inactive foragers are simply the individuals from the total forager population ($n_{foragers}$) that are not recruited for either pollen or nectar foraging:

$$F_{inactive}(t) = n_{foragers} - F_{pollen}(t) - F_{nectar}(t) \qquad (6.3.2.3)$$

6.3.3   Predictions of the Bee Model

Predictions of the bee model show much similarity to the predictions of the wasp and ant models before. At the start of the simulations, the foragers are recruited from the inactive population and they start to fill the common stomachs. The materials (protein and nectar) are used by the colony, and the workforce is balanced quickly to ensure steady operation and material flow. When the colony is perturbed, the workforce changes to compensate for the perturbation (Fig. 6.8). For example, adding proteins to the colony has the effect of decreasing the number of pollen foragers, increasing the number of nectar foragers and also increasing the number of brood. The level of the protein related common stomach (Ω) suddenly increases due to the added protein, but with time it decays back to its previous value. Removing protein from the colony starts an opposite reaction (Fig. 6.8). Perturbations at other points of the model all show similar reactions: namely, resilience and the compensation of workforce to lead the system back to equilibrium (Schmickl and Karsai 2017). We were able to also compare the predictions of our model to data obtained in several classical experiments. For example, Lindauer (1952) used rain machines and pollen traps to study the effect of perturbations on forager activities (Fig. 6.9).

Rain has a serious effect on bee colonies, because the bees cannot forage during the rain. Also, just after the rain, the nectar quality is low, due to a dilution of nectar and because the protein sources inside the colony are probably exhausted. Therefore, it would be adaptive to concentrate on pollen foraging after a rain and switch to a nectar foraging mode later, when the quality of nectar has improved.

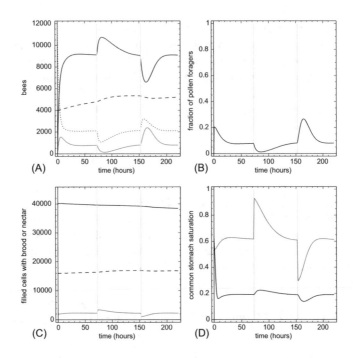

Fig. 6.8 Predictions of the bee model for pollen addition and removal. After about 50 h the system sets an equilibrium independently from the initial conditions. At $t = 72$ h we added 20,000 units of proteins and starting at $t = 152$ h we then removed 20,000 units of proteins within 1 h. The gray vertical lines indicate treatment periods. (**a**) dynamics of task groups and open brood: nectar foragers (black solid line), pollen foragers (gray solid line), inactive foragers (dotted line), and open brood items (dashed line). (**b**) dynamics of the fraction of pollen foragers, calculated as $F_{pollen}(t)/(F_{pollen}(t) + F_{nectar}(t) + 1)$. (**c**) dynamics of pollen stores (gray solid line), nectar stores (solid black line), and sum of open and capped brood (dashed line). (**d**) dynamics of the common stomach saturation: $\Omega(t)$ (gray solid line) and $\Phi(t)$ (black solid line). Reprinted from Schmickl and Karsai (2014) with permission from PLOS

These processes should also depend on the rain patterns. Honeybee colonies operate both in tropical and continental climates. Therefore, their regulation mechanisms should cope with both the rare but long rain periods of continental climate, and with the shorter everyday rains of a tropical climate. We demonstrated that our common stomach model successfully operates in both environments (Fig. 6.10) (Schmickl and Karsai 2016). Rain has a negative effect, because of the cessation of foraging causing some starvation, especially for the larvae, and because the pollen stores are generally quickly exhausted. Why do the bees not store more pollen and prepare better for these rainy periods then?

To test the long-term consequences of this question next, we set up four different strategies for both continental and tropical rain patterns. Not the whole model structure, but only some of the parameters were altered to model the following strategies:

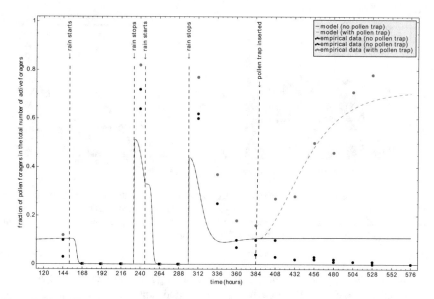

Fig. 6.9 The effect of rain periods and pollen traps on the fraction of pollen foragers. Timing of the manipulations in the model system followed the empirical study of Lindauer (1952): control: black dots; colony with pollen trap (gray dots). Predictions of the model: control: solid line; application of 97% efficient pollen trap: broken line. Reprinted from Schmickl and Karsai (2014) with permission from PLOS

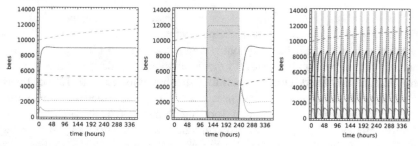

Fig. 6.10 Dynamics of task allocation and brood population. Left panel: control run (no rain). Middle panel: continental rain pattern. Right panel: tropical rain pattern. Gray background areas: rain periods. Solid black line: nectar foragers, solid gray line: pollen foragers, dotted line: inactive foragers, dashed black line: open brood (larvae), dashed gray line: sealed brood (pupae). Reprinted from Schmickl and Karsai (2016) with permission from Elsevier

– Control strategy (no rain),
– Adaptive strategy. This is the strategy that the real honeybees seem to follow as described in our common stomach model.
– Fixed strategy. In this strategy the rain does not play a role, the values of the common stomachs are unchanging (fixed), and therefore the bees forage for pollen with a fixed recruitment rate and do not save proteins. This is a strategy for a minimalistic, non-adaptive constant foraging and expenditure of the pollen.

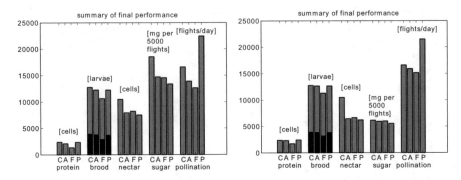

Fig. 6.11 Performance of the modeled honeybee colony in a continental (left) and tropical (right) setting. C: control (no rain) setting. A: "adaptive" strategy; F: "fixed" strategy; P: "proactive" strategy. Brood columns: black indicates open brood, gray indicates capped brood. Reprinted from Schmickl and Karsai (2016) with permission from Elsevier

- Proactive strategy. It is similar to the fixed strategy in that recruitment does not depend on the common stomach, but it is not constant. Here, the pollen foraging rate is increased and kept at a high level until enough pollen is stored that the same number of larvae would survive as in the adaptive strategy. Here the bees stockpile a large quantity of pollen in good weather to avoid starvation in case of rain.

The most significant outcome of these strategies is that the bees had much less sugar gain in the tropics than in the continental setup (Fig. 6.11). This can be explained by the fact that the common rain makes more diluted nectars more frequently. Comparing colony-level fitness parameters in the tropical setting, we observed that the "fixed" strategy leads to higher brood losses (−10.72% compared to the adaptive strategy) while the "proactive" strategy underperforms every other strategy in the sugar economics of foraging flights (−6.67% compared to the adaptive strategy). Also, in the "tropical rain" scenario, the proactive strategy would provide the highest pollination service to the plants. This also means that in the course of plant-pollinator co-evolution, if the plants would be the "driving force," then a proactive strategy would evolve for the bees. This did not seem to happen, however, because this would come with a higher cost of nectar foraging for the proactive colonies. The adaptive strategy (which uses the two common stomachs for regulating task allocation) described in our model is not only in agreement with experimental data, but also provides the best sugar economics— to survive the winter. The adaptive strategy we described in our model is very robust against perturbations and insensitive to initial conditions. We tested this with an extensive sensitivity analysis (again using Vensim's sensitivity tool for 10,000 runs with randomized parameter sets, using a Latin Hypercube Sampling) and as before we did not observe any non-plausible predictions (Fig. 6.12) (Schmickl and Karsai 2016). The common stomach regulation system is also very resilient and compensates well for perturbations.

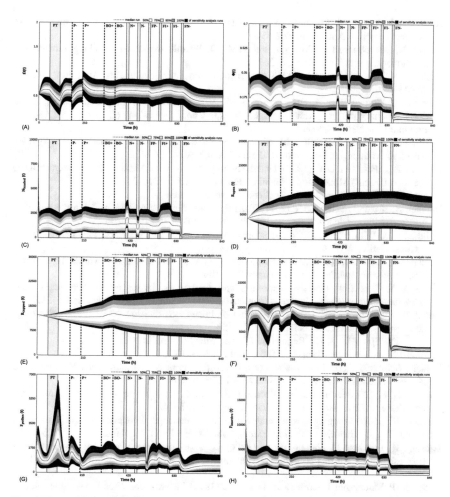

Fig. 6.12 Sensitivity analysis of the major system variables $\Omega(t)$, $\Phi(t)$, and the major stock variables $N_{loaded}(t)$, $B_{open}(t)$, $B_{capped}(t)$, $F_{nectar}(t)$, $F_{pollen}(t)$, and $F_{inactive}(t)$. Dotted lines represent the median values of all runs and bands indicate the deviations of runs from these medians (percentiles and quantiles) of runs: white: 50%, light gray: 75%, dark gray 95%, and black 100% of the 10,000 runs. Perturbations: (+: addition; −: subtraction of materials or entities); PT: pollen trap; P: protein; BO: open larva; N: nectar; FP: pollen forager; FI: inactive forager; FN: nectar forager. Reprinted from Schmickl and Karsai (2014) with permission from PLOS

References

Agbogba C, Howse PE (1992) Division of labour between foraging workers of the ponerine ant *Pachychondyla cafraria* (Smith) (*Hymenoptera; Formicidae*). Insect Soc 39:455–458

Agrawal D, Karsai I (2016) The mechanisms of water exchange: the regulatory roles of multiple interactions in social wasps. PLoS One 11(1):e0145560. https://doi.org/10.1371/journal.pone.0145560PMID:26751076

Anderson C, Franks NR, McShea DW (2001) The complexity and hierarchical structure of tasks in insect societies. Anim Behav 62:634–651

Beckers R, Deneubourg J-L, Goss S, Pasteels JM (1990) Collective decision making through food recruitment. Insect Soc 37(3):258–267

Beshers SN, Fewell JH (2001) Models of division of labor in social insects. Annu Rev Entomol 46:413–440

Blaschon B, Guttenberger H, Hrassnigg N, Crailsheim K (1999) Impact of bad weather on the development of the broodnest and pollen stores in a honeybee colony (*Hymenoptera: Apidae*). Entomol Gener 24(1):49–60

Dussutour A, Latty T, Beekman M, Simpson SJ (2010) Amoeboid organism solves complex nutritional challenges. Proc Natl Acad Sci USA 107(10):4607–4611

Eberlein RL, Peterson DW (1992) Understanding models with Vensim. In: Morecroft JDW, Sterman JD (eds) Modelling for learning. Eur J Oper Res 59:216–219

Frisch KV (1927) Aus dem Leben der Bienen. Springer, Berlin

Hamann H, Karsai I, Schmickl T (2013) Time delay implies cost on task switching: a model to investigate the efficiency of task partitioning. B Math Biol 75:1181–1206

Holldobler B, Wilson EO (2008) The superorganism: the beauty, elegance, and strangeness of insect societies. WW Norton & Company, New York

Huang Z-Y, Otis GW (1991a) Nonrandom visitation of brood cells by worker honey bees (*Hymenoptera: Apidae*). J Insect Behav 4:177–184

Huang Z-Y, Otis GW (1991b) Inspection and feeding of larvae by worker honey bees (*Hymenoptera: Apidae*): effect of starvation and food quantity. J Insect Behav 4:305–317.

Huang Z-Y, Robinson GE (1996) Regulation of honey bee division of labor by colony age demography. Behav Ecol Sociobiol 39:147–158

Karsai I, Phillips MD (2012) Regulation of task differentiation in wasp societies: a bottom-up model of the common stomach. J Theor Biol 294:98–113

Karsai I, Schmickl T (2011) Regulation of task partitioning by a "common stomach": a model of nest construction in social wasps. Behav Ecol 22:819–830

Lachaud JP (1985) Recruitment by selective activation: an archaic type of mass recruitment in a ponerine ant (*Ectatomma ruidum*). Sociobiology 11:133–142

Lachaud JP (1990) Foraging activity and diet in some Neotropical ponerine ants. I. *Ectatomma ruidum Roger* (*Hymenoptera, Formicidae*). Folia Entomol Mex 78:241–256

Lachaud JP, Fresneau D, Garcia-Perez J (1984) Etude des strategies d'approvisionnement chez trois especes de fourmis ponerines (Hymenoptera, Formicidae). Folia Entomol Mex 61:159–177

Lachaud JP, Valenzuela J, Corbara B, Dejean A (1990) La predation chez Ectatomma ruidum: etude de quelques parametres environnementaux. Act Colloq Insect S 6:151–155

Lindauer M (1952) Ein Beitrag zur Frage der Arbeitsteilung im Bienenstaat. Z Vergl Physiol 34:299–345

Middleton EJT, Latty T (2016) Resilience in social insect infrastructure systems. J R Soc Interface 13:2015–1022

Plowright RC, Plowright CMS (1988) Elitism in social insects: a positive feedback model. In: Jeanne RL (ed) Interindividual behavioral variability in social insects. Westview Press, Boulder pp 419–431

Pratt SC, Mallon EB, Sumpter DJT, Franks NR (2002) Quorum sensing, recruitment, and collective decision making during colony emigration by the ant *Leptothorax albipennis*. Behav Ecol Sociobiol 52:117–127

Riessberger U, Crailsheim K (1997) Short-term effect of different weather conditions upon the behaviour of forager and nurse honey bees (*Apis mellifera carnica Pollmann*). Apidologie 28(6):411–426

Robinson GE (1992) Regulation of division of labor in insect societies. Annu Rev Entomol 37:637–665

Schatz B, Bonabeau E, Theraulaz G, Deneubourg JL (1999a) Modèle de division du travail bas è sur des seuils de réponse chez une fourmi ponérine. Act Colloq Insect S 12:19–22

Schatz B, Lachaud JP, Beugnon G, Dejean A (1999b) Prey Density and Polyethism within Hunting Workers in the Neotropical Ponerine Ant *Ectatomma ruidum* (*Hymenoptera, Formicidae*). Sociobiology 34:605–617

Schatz B, Lachaud JP, Beugnon G (1995) Spatial fidelity and individual foraging specializations in the neotropical ponerine ant, *Ectatomma ruidum Roger* (*Hymenoptera; Formicidae*). Sociobiology 26:269–28

Schatz B, Lachaud JP, Beugnon G (1997) Graded recruitment and hunting strategies linked to prey weight and size in the ponerine ant *Ectatomma ruidum*. Behav Ecol Sociobiol 40:337–349

Schmickl T, Blaschon B, Gurmann B, Crailsheim K (2003) Collective and individual nursing investment in the queen and in young and old honeybee larvae during foraging and non-foraging periods. Insect Soc 50:174–184

Schmickl T, Crailsheim K (2001) Cannibalism and early capping: strategy of honeybee colonies in times of experimental pollen shortages. J Comp Physiol A 187:541–547

Schmickl T, Crailsheim K (2002) How honeybees (*Apis mellifera* L.) change their broodcare behaviour in response to non-foraging conditions and poor pollen conditions. Behav Ecol Sociobiol 51:415–425

Schmickl T, Karsai I (2014) Sting, carry and stock: how corpse availability can regulate decentralized task allocation in a ponerine ant colony. PLoS One 9(12):e114611. https://doi.org/10.1371/journal.pone.0114611

Schmickl T, Karsai I (2016) How regulation based on a common stomach leads to economic optimization of honeybee foraging. J Theor Biol 389:274–286

Schmickl T, Karsai I (2017) Resilience of honeybee colonies via common stomach: a model of self-regulation of foraging. PLoS One 12(11):e0188004. https://doi.org/10.1371/journal.pone.0188004

Seeley TD (1982) Adaptive significance of the age polyethism schedule in honeybee colonies. Behav Ecol Sociobiol 11:287–293

Seeley TD (1992) The tremble dance of the honey bee: message and meanings. Behav Ecol Sociobiol 31:375–383

Seeley TD, Camazine S, Sneyd J (1991) Collective decision-making in honey bees: how colonies choose among nectar sources. Behav Ecol Sociobiol 28(4):277–290

Theraulaz G, Bonabeau E, Sole RV, Schatz B, Deneubourg JL (2002) Task partitioning in a ponerine ant. J Theor Biol 215:481–489

Theraulaz G, Bonebeau E, Deneubourg JL (1998) Response threshold reinforcement and division of labour in insect societies. Proc R Soc Lond B 265:327–332

Wilson EO (1971) The insect societies. Belknap Press of Harvard University Press, Cambridge

Chapter 7
Generalization of the Common Stomach: Integral Control at the Supra-Individual Level

Abstract Homeostatic self-regulation is an essential aspect of open dissipative systems presented in the previous chapters. These systems show strong resilience against perturbations and are regulated by interactions networks. In this chapter, we generalize our findings and present the regulation role of redundant (predominantly, negative) feedback systems paired with a buffer node. We present this regulation system as a type of integral control which is responsible for the robustness of many homeostatic mechanisms. The regulation core, which we call common stomach regulation, describes close regulatory relationships between a substance and the colony members that process that substance. In some sense, the substance itself is regulating its collection and use by "adjusting" the workforce that handles that material. Robust systems have the important advantage of having a larger parameter space, in which they can evolve and adjust to environmental changes. Identifying and understanding the nature of control mechanisms are of fundamental importance for the understanding of biological regulation. The common stomach regulatory network we discovered is an example of closed loop systems that work at the supra-individual level. Different insect societies evolved their eusociality independently from non-social predecessors, but they have "discovered" the same, very efficient regulatory mechanism.

7.1 Background

Homeostatic self-regulation is an essential aspect of open dissipative systems, such as the majority of biological systems. In the earlier chapters we have presented examples for ecological systems that showed resilience against perturbations. They were regulated by interactions of individuals with each other and the environment. In the last chapters we presented social systems (insect societies) that were also very resilient yet adaptive, where material flow was regulated by the material itself (the saturation of the common stomach) through an interaction of the material and the workforce. Although the term "homeostasis" was originally used to refer to processes within living organisms, today it is applied in many fields. For example, engineering systems have recently begun to have biological levels of complexity and

© Springer Nature Switzerland AG 2020
I. Karsai et al., *Resilience and Stability of Ecological and Social Systems*,
https://doi.org/10.1007/978-3-030-54560-4_7

the planning of better machines could imply applying ideas from biology. Analyses of both biological and physical systems show that protocols and regulatory feedback loops that ensure optimality and robustness are the most important components to biological complexity (Wiener 1948; Csete and Doyle 2002; Åström and Kumar 2014). In engineering, these protocols need to be designed, but in biological systems they have evolved spontaneously. Here we can expect that successful protocols become highly conserved (and thus general) because they facilitate evolution and are difficult to replace.

Generally, negative feedback is associated with homeostasis, but complex systems always have a redundant feedback network instead of a single feedback loop in place. (Yi et al. 2000) have argued that robust asymptotic tracking requires some kind of integral feedback as a structural property of the system. More generally, integral control may underlie the robustness of many homeostatic mechanisms. Integral feedback is, for example, used commonly in engineering, and it is likely to be common in biology as well, in achieving homeostatic regulation or "perfect adaptation" (Yi et al. 2000; Csete and Doyle 2002).

Integral feedback control is a fundamental engineering strategy for ensuring that the output of a system robustly sets to an equilibrium value that is resilient to noise or a variation in the system's parameters. Analyzing challenges and constraints arising in the efforts to engineer biological integral feedback controllers has shown that resource limitations are key to understand these controllers. Resource limitations restrict the amount by which gains can be increased, and this, along with other physical constraints, affect the feedback's design and function (Yi et al. 2000; Ang and McMillen 2013; Del Vecchio et al. 2016).

For example, saturation acting as a control for integral feedback is a key property for many negative feedback mechanisms (Ang and McMillen 2013). Recognizing integral feedback control in biological systems is important to biologists, because these could provide a mechanistic explanation of many biological phenomena.

The robust ability of a system to adapt quickly stems predominantly from the network connectivity that exists between the components of the system, without requiring a distinctive fine-tuning of system parameters (Berg and Tedesco 1975; Koshland et al. 1982; Khan et al. 1995). Even the simplest networks can have large variety of different configurations. Ma et al. (2009) have computationally investigated all possible three-node network topologies, to identify those that could perform adaptation (i.e., reach a homeostatic value). Only two major topologies showed a high degree of robustness: A negative feedback loop with a buffering node and an incoherent feed-forward loop with a "proportioner node." They concluded that, within proper regions of the parameter space, minimal interaction networks containing these topologies are sufficient to achieve adaptation. More complex networks that robustly perform adaptation should also contain at least one of these topologies at their core. It seems that negative feedback alone is not capable of robustly pushing the system to equilibrium. Ma et al. (2009) found that different negative feedback loops differ widely in their ability to facilitate adaptation. There is only one class of simple negative feedback loops that can robustly achieve adaptation: it is when the output node does not feed directly back to the input node, but instead goes through an intermediate node that serves as a buffer. This is a key property of the regulation loops we found in insect societies and named as the common stomach regulation system.

7.2   Insect Societies and the Common Stomach

Insect societies depend on coordinated complex infrastructure systems, such as supply chains, transportation and communication networks, and storage. How these complex systems are regulated is still not completely understood. The queen is not in command and there are no foremen. So much is sure. These systems have a decentralized control, where individual insects make simple decisions based on local information (Gordon 2016; Crall et al. 2018). These societies also show both a high level of adaptability to changes and a strong resilience against perturbations (O'Donnell and Bulova 2007). In the previous chapters we have discussed the task allocation mechanisms of wasps (Karsai and Wenzel 1998, 2000; Karsai and Schmickl 2011; Karsai and Phillips 2012), ants (Schmickl and Karsai 2014), and honey bees (Schmickl and Karsai 2016, 2017). In all these cases, we have demonstrated that the workers were able to switch tasks without changes in their development or a learning experience. The switches between tasks were quick or caused short-term time delays (Hamann et al. 2013) and were reversible. These switches result in rapid adjustment of the workforce, allowing the colony to compensate for disturbances or changes in demands.

In each studied case, we were able to identify a given species-specific substance that act as the key for the regulation. Under the term "substance" we do not mean pheromones, which are commonly mentioned in insect societies—such as trail pheromones in the ants or the specific pheromone of the queen bee. Rather, these substances in question are materials used by the colony for something more than to merely convey information. These substances are stored in a saturable "common stomach" and they are removed from this storage when the colony uses them. The individual workers are interacting with this storage, which also acts as an information center and as a buffer. These common stomachs store and regulate the inflow and outflow of substances, and the saturation of these substances in the common stomachs regulates the task allocation of material gatherers and users. In general, the local density of a substance corresponds to the global system state, if integrated over time. This means that a simple cue-based regulation can suffice to regulate global system states without the need to develop a more complex signal- or language-based communication. By simplifying the regulation systems as described in the 3 different insect societies (wasps, ants, and bees), we found a number of similarities (Fig. 7.1).

The regulation scheme is distributed over three layers. The first layer corresponds to an open system describing the inflows and outflows of the substances in the colony. The amount of these substances and the storing capacity of the colony

Fig. 7.1 The three layers of the regulation systems of task allocation and the material flows in three insect societies. Substance: yellow; workforce: red; information layer: blue. The regulating key variables in each insect society are the stocks of shared substances (common stomach) and the saturation of those substances (expressed by Greek letters in circles). Recruitment and abandonment of individuals to and from the different task categories is regulated by the common stomach saturation, which in turn, affects the material flows. Reprinted from Schmickl and Karsai (2018) with permission from the National Academy of Sciences

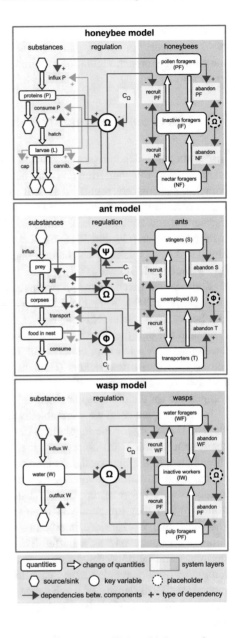

together form a degree of saturation of the common stomach(s), which next forms the central information layer. The last layer represents the workforce, and this, for simplicity, is assumed to be a simple closed system. Workers in the workforce pool can change from an inactive worker to either a substance provider or a substance user or the other way around. These task changes are governed by the interactions of the middle information layer, namely the saturation of the common stomach. The size of workforce will determine in every task group how much substance will be collected, stored, and used and this, in turn, determines the saturation of the common stomach.

7.3   Model of the Common Core of the "Common Stomach" Regulation Mechanism

Our studies suggest that there is a commonality in the task regulation mechanism of the different insect societies. We assumed that there is a core regulation system. We anticipated that it is possible to construct a function-topology mapping that captures the essential topologies underlying the task allocation and material flows observed. We proposed that the fundamental principle of the task allocation mechanism is akin to integral control in insect societies (Fig. 7.2). When the saturation of the common stomach is low, more foragers are recruited from unemployed workers, while some of the consumers will become inactive (unemployed). A consequence of this task reallocation process will be an increased inflow and a reduced outflow of the substance in question and this leads to an increase of the saturation level of the common stomach. If the common stomach is highly saturated, this indicates that a lot of substance is available to the colony. This, in turn, encourages the recruitment of more consumers and the abandonment of foraging.

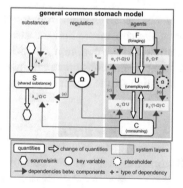

Fig. 7.2 Schematic representation of the core regulatory mechanism of insect societies. Substance (S) is used for one or several important colony functions. The saturation level (Ω) of S affecting recruitments and abandonments of substance users and foragers and the rate of usage. Arrows with blue letters (a–e) indicate causal connections where the common stomach saturation (Ω) affects other components directly. Gray boxes describe the terms of the model. Reprinted from Schmickl and Karsai (2018) with permission from the National Academy of Sciences

The model of the core of the common stomach regulation system is very simple. It is available for downloading.[1]

We assume that the change of shared substance $S(t)$ in the colony depends on an inflow and an outflow term:

$$\frac{dS}{dt} = \lambda_{in} F(t) - \lambda_{out} \Omega(t) C(t) \tag{7.3.1}$$

The foragers $F(t)$ will bring a volume λ_{in} of substance into the nest and this substance is consumed by the consumers $C(t)$ with a rate λ_{out} multiplied by the saturation level $\Omega(t)$ of the common stomach. This simply means that if there is more substance in the colony, the consumption becomes higher and when the substance level is low, the consumption decreases. The change in the number of foragers (F) and consumers (C) is determined by a recruitment process—unemployed workers turn to special task (first terms) and an abandonment process (turns special task into unemployed (second terms):

$$\frac{dF}{dt} = \alpha_F (1 - \Omega(t)) U(t) - \beta_F \Omega(t) F(t) \tag{7.3.2}$$

$$\frac{dC}{dt} = \alpha_C \Omega(t) U(t) - (1 - \beta_C \Omega(t) C(t)) \tag{7.3.3}$$

The recruitment of foragers and consumers happens from the unemployed worker category $U(t)$ and is scaled by the recruitment coefficients (α-s). Abandonment turns specialized task workers ($F(t)$ and $C(t)$) into unemployed, scaled by abandonment rates (β-s). Each term also receives input from the saturation of the common stomach: the recruitment of the consumers and the abandonment of the foragers are proportional to the saturation of the common stomach, while the recruitment of foragers and the abandonment of the consumers are inversely proportional to the common stomach. The saturation of the common stomach is calculated from the quantity of substance:

$$\Omega(t) = S(t)/s_{max} \tag{7.3.4}$$

where s_{max} is the overall storage capacity of the common stomach, which depends of the number of entities which engage in the processing of the substance (foragers, users, and storage units) and the storage capacity of these entities. For simplicity, we assumed that this maximum capacity is a constant. We further assumed that the number of workers participating in the process (N) is also constant, therefore, the rates of change of the unemployed workers can be calculated directly as $U(t) = N - C(t)+) + F(t)$.

We can conclude that this regulatory mechanism is akin to the well-known integral feedback mechanisms described in many physical and biological systems

---

[1] https://sites.google.com/site/springerbook2020/chapter-7.

at or below the cellular levels. We showed that the core of the common stomach regulation is based primarily on negative feedback where the output indirectly feeds back to the input through an intermediate node that also serves as a buffer. Here we identified an integral feedback regulatory mechanism, but one which works at the organizational level of animal societies. We emphasize that this is an important core mechanism that ensures the resilience and stability of insect societies.

7.4 Properties of the Common Core Regulation Mechanism

The robustness and the insensitivity to initial conditions of the common stomach regulation system stem from the redundant feedback network. We found 19 feedback loops (Fig. 7.2) of which 13 are negative, thus stabilizing, and six are positive, thus escalating, feedback loops (Table 7.1). This redundancy increases the reliability of the regulation system. To demonstrate this fact and to assess the importance of the major feedback loops, we started to systematically cut the links (Fig. 7.2, arrows with letters a–e) that connected the saturation of the common stomach (Ω) to other system components. We initialized a normal run and after 20 steps we experimentally induced a sudden leaking flow of the common stomach for 15 steps then we closed this leak again. Generally, the system shows a quick convergence towards equilibrium, a robust counterbalancing of the perturbation, and a fast return to the original equilibrium after the leak of the common stomach is closed (Fig. 7.3). Depending on what feedback is severed, however, the colony reacts somewhat differently. For example, the elimination of the feedback link to the outflow of the substance (Fig. 7.3, black line) results in increased fluctuations and stronger reactions to the perturbation. Eliminating the feedback link to the recruitment of both worker tasks has a similar, but weaker effect (Fig. 7.3, white line). Cutting the feedback link to the abandonment of both tasks causes a quicker reach of the equilibrium, but a decreased ability to compensate for the effect of a perturbation (Fig. 7.3, green broken line). Abandoning tasks is very important to ensure that there will be a large enough workforce to be recruited from for coping with an emergency. Thus, when we sever the feedback from the common stomach saturation to the abandonments, the consumers cannot abandon consumption quickly, when, e.g., a substance leak is implemented. Therefore, they continue consuming and decrease the already scarce substance further, making the situation even worse. However, even in this case, the system will still work and not break completely. We concluded that the minimum regulation of a working system occurs with feedback links of the common stomach saturation to substance outflow and to the recruitment of foragers (Fig. 7.3, gray line).

It is important that colonies can react to the changes of internal needs and major environmental perturbations. However, it is also important not to react to the smaller fluctuations, generally described as random noise. For example, if the recruitment or abandonment of special tasks would react to any small changes in the substance inflow or usage, then the wasps would too frequently change tasks. The common

Fig. 7.3 The importance of main feedback loops and the dynamics of foragers (F), consumers (C), and common stomach saturation (Ω). Specific feedback components (Fig. 7.2, arrows with letters a–e) are either turned on ($+$) or off ($-$) for different runs (visualized by line types). Between 20 and 35 units of time, a 20% leakage on the substance stock is implemented. Broken black line: normal run with all feedbacks active; black solid line: feedback to substance consumption cut; white solid line: feedback to recruitments cut; green broken line: feedback to abandonments cut; gray line: minimum required feedback, necessary for the regulation still operable. Reprinted from Schmickl and Karsai (2018) with a permission from the National Academy of Sciences

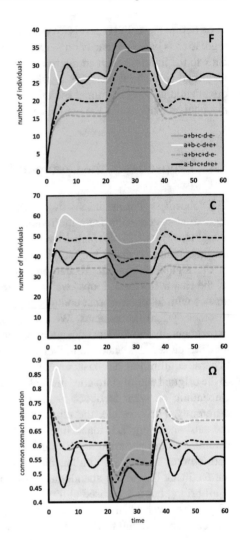

stomach thus has a very important property: it is also a buffer. The saturation of the common stomach changes more slowly than the substance in- and outflows.

For example, the environmental influx of shared substance, along with the colony's internal usage of the substance, can vary throughout the day (Dussutour et al. 2009; Deneubourg et al. 1983). Variations in the influx may stem from the stochasticity of substance retrieval (the scouting success in honeybees, Seeley 1983). Crop-loads returned by foragers vary individually and they depend on environmental factors (Huang and Seeley 2003; Li et al. 2014). To investigate how the dynamic equilibria of the common stomach regulation system are affected by varying levels of noise, we performed a set of targeted simulation experiments, where one of the flux rates was additively transformed by normally distributed

Table 7.1 Identification of feedback loops of the common stomach regulation system. Reprinted from Schmickl and Karsai (2018) with permission from the National Academy of Sciences

Loop	Variables included in each feedback loop	Type	Effect
#1	$S(t) \rightarrow \Omega(t) \rightarrow outflux(t) \rightarrow S(t)$	Negative	Prevents CS from running empty
#2	$U(t) \rightarrow recruitment_F(t) \rightarrow U(t)$	Negative	Balances the number of unemployed agents
#3	$U(t) \rightarrow recruitment_C(t) \rightarrow U(t)$	Negative	Balances the number of unemployed agents
#4	$F(t) \rightarrow abandonment_F(t) \rightarrow F(t)$	Negative	Balances the number of active foragers
#5	$C(t) \rightarrow abandonment_C(t) \rightarrow C(t)$	Negative	Balances the number of active consumers
#6	$S(t) \rightarrow \Omega(t) \rightarrow abandonment_C(t) \rightarrow C(t) \rightarrow outflux(t) \rightarrow S(t)$	Negative	Prevents the CS from running empty
#7	$S(t) \rightarrow \Omega(t) \rightarrow abandonment_F(t) \rightarrow F(t) \rightarrow influx(t) \rightarrow S(t)$	Negative	Prevents the CS from overfilling
#8	$S(t) \rightarrow \Omega(t) \rightarrow recruitment_C(t) \rightarrow C(t) \rightarrow outflux(t) \rightarrow S(t)$	Negative	Prevents the CS from running empty
#9	$S(t) \rightarrow \Omega(t) \rightarrow recruitment_F(t) \rightarrow F(t) \rightarrow outflux(t) \rightarrow S(t)$	Negative	Prevents the CS from overfilling
#10	$U(t) \rightarrow recruitment_F(t) \rightarrow F(t) \rightarrow abandonment_F(t) \rightarrow U(t)$	Positive	Enhancing foraging
#11	$U(t) \rightarrow recruitment_C(t) \rightarrow C(t) \rightarrow abandonment_C(t) \rightarrow U(t)$	Positive	Enhancing consumption
#12	$U(t) \rightarrow recruitment_F(t) \rightarrow F(t) \rightarrow influx(t) \rightarrow S(t) \rightarrow \Omega(t) \rightarrow recruitment_C(t) \rightarrow U(t)$	Negative	Balancing the tasks
#13	$U(t) \rightarrow recruitment_C(t) \rightarrow outflux(t) \rightarrow S(t) \rightarrow \Omega(t) \rightarrow abandonment_C(t) \rightarrow U(t)$	Positive	Enhancing consumption
#14	$U(t) \rightarrow recruitment_C(t) \rightarrow outflux(t) \rightarrow S(t) \rightarrow \Omega(t) \rightarrow recruitment_F(t) \rightarrow U(t)$	Negative	Balancing the tasks
#15	$U(t) \rightarrow recruitment_F(t) \rightarrow influx(t) \rightarrow S(t) \rightarrow \Omega(t) \rightarrow abandonment_F(t) \rightarrow U(t)$	Positive	Enhancing influx
#16	$U(t) \rightarrow recruitment_C(t) \rightarrow outflux(t) \rightarrow S(t) \rightarrow \Omega(t) \rightarrow abandonment_F(t) \rightarrow U(t)$	Negative	Regulating outflux
#17	$U(t) \rightarrow recruitment_F(t) \rightarrow influx(t) \rightarrow S(t) \rightarrow \Omega(t) \rightarrow abandonment_C(t) \rightarrow U(t)$	Negative	Regulating influx
#18	$U(t) \rightarrow recruitment_F(t) \rightarrow influx(t) \rightarrow S(t) \rightarrow \Omega(t) \rightarrow recruitment_F(t) \rightarrow C(t) \rightarrow abandonment_C(t) \rightarrow U(t)$	Positive	Enhancing influx
#19	$U(t) \rightarrow recruitment_C(t) \rightarrow outflux(t) \rightarrow S(t) \rightarrow \Omega(t) \rightarrow recruitment_F(t) \rightarrow F(t) \rightarrow abandonment_F(t) \rightarrow U(t)$	Positive	Enhancing consumption

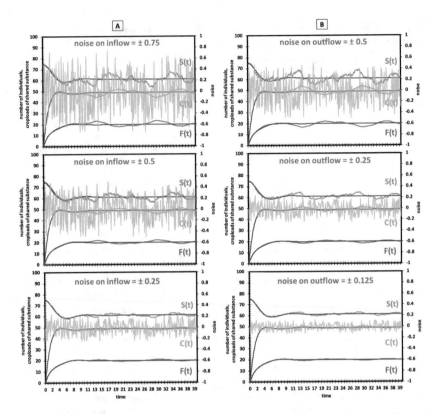

Fig. 7.4 Effect of zero-mean noise (gray) in the substance inflow (**A**) and outflow (**B**) on the dynamic equilibria of the system variables $S(t)$: amount of substance; $C(t)$: number of consumers; $F(t)$: number of foragers. Reprinted from Schmickl and Karsai (2018) with a permission from the National Academy of Sciences

zero-mean noise (non-proportional transformation). In each time step, a randomly generated value (between zero and the noise level 0.0, 0.25, 0.5, 0.75 and a standard deviation of 66.6% of the noise level, taken from a normal distribution) was added or subtracted to the influx or outflux value. We observed that the system showed a high-level stability against different levels of zero-mean noise, and the main system parameters remained close to their equilibrium values (Fig. 7.4). The fluctuations are strongest in the shared substance depot $S(t)$. However, this could be expected, as the common stomach is the integrator of these flows. Integrators, or buffers in general, tend to decrease the impact of high-frequency system noise (Peng et al. 2009). Due to the buffering effect of the common stomach, the compositions of workers are less affected by the noise and therefore the common task switches are avoided.

In the insect societies not only do the inflow and outflow of substance vary, but there is some variance between the individuals as well. Much as we did for the models of the three different societies in the previous chapters, we have carried out a set of extensive sensitivity analyses for the core common stomach model here. Pairing perturbation experiments with an extensive parameter sweeping has showed that the system is very robust. The common stomach regulation mechanism compensated for the strong perturbations by keeping the amount of substance (S) and the saturation level of the common stomach (Ω) very stable, while counteracting the perturbations by rearranging the workforce (Fig. 7.5). The system is also very resilient—after every perturbation, it re-establishes its equilibrium very quickly.

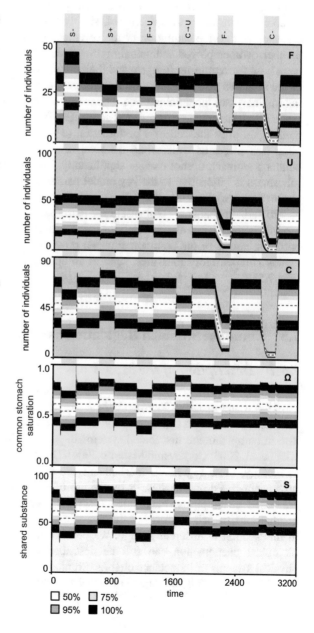

Fig. 7.5 Sensitivity analysis of the core model of the common stomach regulation system. Panels: F: foragers; U: Unemployed; C: consumers; S: amount of substance; Ω: saturation of the common stomach. Perturbations (vertical gray columns): removing substance ($S-$); adding substance ($S+$); 25% of foragers turned to unemployed workers ($F \rightarrow U$); all consumers turned to unemployed workers ($C \rightarrow U$): removal of foragers from the colony ($F-$); removal of consumers from the colony ($C-$). The broken blue line indicates the median of the runs; a white ribbon indicates 50% of results; the light and dark gray ribbons indicate 75% and 95% of the results, respectively; and a black ribbon indicates 100% of the results of 10,000 simulations (using Vensim's sensitivity tool Eberlein and Peterson 1992), Latin Hypercube Design. Reprinted from Schmickl and Karsai (2018) with a permission from the National Academy of Sciences

Removing $(S-)$ or adding $(S+)$ material to the substance stock was compensated for quickly via recruiting more foragers or consumers, respectively (Fig. 7.5). Similarly, experimentally transferring some individuals from one task group to another $(F \rightarrow U, C \rightarrow U)$ was compensated by stronger recruitments for the missing members of the task groups and an increased abandonment from the task groups, with a surplus of workers. The dynamics of how these trends happened were more complex.

For example, when foragers are forced to turn into unemployed workers, the number of consumers also decreases, simply because relatively less substance is collected and so the saturation of the common stomach decreases. Therefore, the number of unemployed individuals grows even further. This, in turn, provides a large base for recruiting foragers and the system re-establishes itself quickly. Most of the time this also happens with a slight over-recruitment of the foragers, which, in turn, increases the number of consumers, because the common stomach becomes better filled. Removing foragers and consumers from the colony and then returning them again after the end of a perturbation $(F-, C-)$ result in a decrease in the other task forces as well, while the level of substance and the saturation of the common stomach do not change significantly. The system is very robust and this robustness is insensitive to the key model parameters (the α, β, and λ values) and also to the type of perturbations. The common stomach regulation system is able to absorb sudden changes, compensate for perturbations—and the simulations never resulted in any "nonsense" predictions (such as a negative number of workers). The buffering nature of the common stomach provides enough time for the colony to deploy compensatory measures, such as an appropriate shifting of the workforce (Fig. 7.5).

7.5 Common Stomach Regulation as a Circuit Model

7.5.1 Description of the Model

Workforce and material flow regulations are general challenges not only in insect societies (Deneubourg and Goss 1989; Seeley et al. 1991; Schmickl et al. 2012), but also in computational distributed systems in general (Bannister and Trivedi 1983; Klügl et al. 2003; Gerkey and Matari'c 2004). Due to the lack of "master equations" in biology, these regulation mechanisms are commonly described and modeled with agent-based approaches, or by the use of empirical functions. These models are based on empirical data, fitted functions, and some simple and reasonable assumptions, which can predict the operation of natural colonies well. Our goal in this section is different though: we present how the essence of this complex biological phenomenon can still be described by master equations, using the physical systems of inductance circuits. Similar models based on electrical circuits have been successfully attempted to model such systems as the nervous system. For

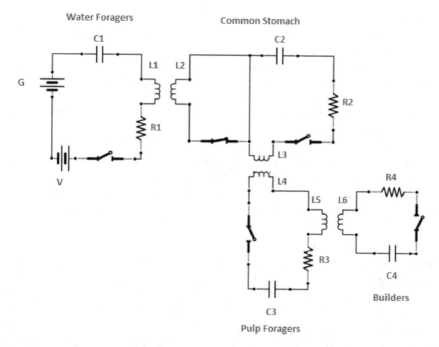

Fig. 7.6 Circuit diagram model of task partitioning in *Metapolybia* wasps. The four circuits represent the groups of wasps belonging to the four respective task groups. Reprinted from Hilbun and Karsai (2016) with a permission from PLOS

example, Hodgkin and Huxley (1952) provided a circuit model, based upon resistors and a capacitor, to model nerve impulses.

We have built a theoretical model from electric circuits that can provide predictions similar to the data in real colonies and the predictions of empirical models. We only describe the skeleton of the model here. For details see Hilbun and Karsai (2016) and the published demonstration.[2]

Coupled circuits, involving capacitors, resistors, and inductors were chosen for this model due to the circuits' inherent abilities to essentially explain storage ability via inductors and capacitors, loss caused by environmental factors via resistors, as well as having a general structure that would allow for a circular flow of a supply of particles: water in the biological system and electrons in the physical circuit, respectively. The model uses the terminology of wasp societies we described in Chap. 5. The model consists of four interconnected electrical circuits, each of which acts as a different functional part of the wasp colony: water foragers, the common stomach, pulp foragers, and builders (Fig. 7.6). Electrons flowing through

---

[2]http://demonstrations.wolfram.com/TaskPartitioningInInsectSocietiesAModelBasedOnElectric Circuits/.

the circuits are used to model the water flow through the system. The electron is the substance in this model, that is collected, used by the elements of the circuits and these form the substance that can saturate the common stomach. The RLC circuits we used are related to simple harmonic oscillators. Inductance (L) acts as the mass of a harmonic oscillator system; resistance (R) causes damping, and capacitance (C) behaves like the spring constant of a mass, oscillating on a spring. Each of the individual RLC circuits is connected by mutual inductance, representing the connectedness of the tasks.

We used mutual inductance as a connection between the circuits because the change of mutual inductance in one coil induces a current in the second coil and this is similar to how task groups are interacting. The water foragers acquire water, and then the water foragers directly affect the common stomach, in turn, the common stomach then directly affects the pulp foragers, and the pulp foragers directly affect the builders. Sinks of water for construction and drinking/cooling are modeled via resistors.

Water Foragers The task of the water foragers is to collect water and transfer it to the common stomach. A circuit, which can model these water foragers comprises a capacitor ($C1$), a resistor ($R1$), and an inductor ($L1$). The circuit also has two voltage sources, (V and G), allowing current to flow in two different directions and for directionality to be adjusted. The water forager circuit's inductor ($L1$) is placed in close proximity to the inductor ($L2$) of the common stomach to allow for a flow between the two circuits. Water (viz. electrons) for the whole system is generated by G.

Common Stomach The common stomach's RLC circuit is comprised of a resistor ($R2$), a capacitor ($C2$), and two inductors ($L2$ and $L3$). The function of the common stomach is to store, buffer, and connect the water forager and the user task groups. This flow of materials is modeled by $L1$ and $L2$ and $L3$ and $L4$, the inductors being placed in close proximity. There is a wire which divides the circuit so that this circuit will have two switches, each affecting a different inductor, allowing for feedback. One switch opens when the other switch closes, and vice versa. This causes changes in the magnetic field so that voltage can be induced in adjacent circuits (by Faraday's Law). Additionally, because the common stomach is a temporal storage place, it has a high capacitance capability. The resistor ($R2$) of the common stomach simulates the leakage of the common stomach (i.e., water used for other non-construction related tasks such as cooling). The central location of this circuit allows it to act as a buffer and also provide feedback to the system.

Pulp Foragers The pulp forager RLC circuit is a combination of a capacitor ($C3$), a resistor ($R3$), and two inductors ($L4$ and $L5$) serially connected, and it is attached to the common stomach via the inductor $L4$ and to the builders via $L5$. To simplify the system, we hereby assumed that the pulp foragers simply convert water to watery pulp, therefore water is lost only in small quantities through $R3$ (i.e., some water evaporates during pulp making).

Builders The builders' RLC circuit is a combination of a capacitor ($C4$), a resistor ($R4$), and an inductor ($L6$) serially connected, and it is attached to the pulp foragers via the inductor $L5$. The main sink of electrons is modeled by a resistor ($R4$). The conversion of electrons into heat represents water loss during the construction: water evaporates from the nest when the builders apply watery pulp to the nest.

After constructing these circuits, we used the master equations of Kirchoff's fundamental laws to model the dynamics of the four components. The calculated charges correspond to the quantities of workers and materials described before using the empirical models. In this circuit model, the change of charge over time in the water forager circuit, i.e. $W''(t)$, is described by

$$W''(t) = \frac{C''(t)M1 - R1W'(t) - \frac{W(t)}{C1} + V_{Battery}}{L1} \tag{7.5.1.1}$$

here the $C''(t) * M1$ term represents the mutual inductance term of water foragers connected to the common stomach, $R1 * W'(t)$ is voltage drop due to the resistor representing water use of the water foragers, and $W(t)/C1$ is subtracted as the voltage drop across the capacitor, showing the water foragers' ability to retain small quantities of water. The water inflow is then modeled via the battery voltages, V and G. These are summed and together referred to as $V_{Battery}$. The right-hand side of the equation is divided by $L1$, which was derived as $W''(t)$ multiplied by $L1$ is the change of current times the inductance of the circuit, also caused by mutual inductance. The equation is set equal to zero, and then it can be solved for $W''(t)$, causing all terms to be divided by the $L1$ inductor. We can use the same logic to do the other 3 equations as well.

The change of charge in the common stomach circuit $C''(t)$ is given by

$$C''(t) = \frac{W''(t)M1 - P''(t)M2 - R2C'(t) - \frac{C(t)}{C2}}{L2 + L3} \tag{7.5.1.2}$$

where $W''(t) * M1$ represents the first mutual inductance term of the water foragers acting with the common stomach, $P''(t) * M2$ is the second mutual inductance term of the common stomach, creating mutual inductance with the pulp foragers. These two terms allow for the transfer of water to the common stomach by the water foragers and from the common stomach by the pulp foragers. An $R2 * C'(t)$ term is subtracted as the voltage drops across this resistor (leak of the common stomach). Also a $C(t)/C2$ term is subtracted for the large capacitor, which plays the role of water storage, buffer and eventually regulating the wasp activity. The right-hand side of the equation is rearranged in the same way as described for the water forager above. The change of charge for the pulp forager circuit $P''(t)$ is modeled as:

$$P''(t) = \frac{C''(t)M2 - B''(t)M2 - R3P'(t) - \frac{P(t)}{C3}}{L4 + L5} \tag{7.5.1.3}$$

here $C''(t) * M2$ represents the mutual inductance term of the pulp foragers, connected with the common stomach, and $B''(t) * M2$ is the mutual inductance term of the pulp foragers with the builders. These two terms allow for the transfer of water from the common stomach to the pulp foragers and also from the pulp foragers to the builders. $R3 * P'(t)$ is simply the voltage drop across the resistor from Ohm's Law, showing water use (other than pulp collecting behavior) of the pulp foragers. $P(t)/C3$ is subtracted as the voltage drop across the capacitor, showing the pulp foragers' ability to store small quantities of water. The change of charge in the builder circuit $B''(t)$ is described by

$$B''(t) = \frac{P''(t)M2 - R4B'(t) - \frac{B(t)}{C4}}{L6} \qquad (7.5.1.4)$$

where $P''(t) * M2$ represents the mutual inductance term of the pulp foragers acting with the builders. This models the water arriving to the builders in the form of a wet pulp. An $R4 * B'(t)$ term is subtracted as the voltage drop across this resistor, modeling the evaporating water leaving the freshly constructed nest material (main sink of the system). A $B(t)/C4$ term is subtracted for the capacitor showing the builders' ability to store small quantities of water. As for the other equations above, the numerator on the right-hand side of the equation is divided by the inductance term, because the derivation was that $B''(t) * L6$ represents the second aspect of the mutual inductance, which is dependent upon the inductance of the individual coils that are in the common stomach coil. The equation is then solved for $B''(t)$, so the right-hand side of the equation must be divided by $L6$.

7.5.2 Predictions of the Circuit Model

The model based on the master equations of electric circuits is able to mimic the experimental data observed on wasps (Karsai and Wenzel 2000), as well as the predictions of both our top-down and agent-based models (see Chap. 5). The system reaches an equilibrium, while electrons are moving from the generator (the main electron source) to the resistor of the builder circuit (the main electron sink). The system reacts in a similar way to the experimental perturbations as observed before. For example, to simulate the capturing of water foragers (i.e., removing some members of this task group), we suddenly decreased their number by reducing $C1$ from 0.5 to $0.25F$ and reducing $R1$ from 2 to 1Ω. This reduced number of water foragers is unable to refill the common stomach, which in turn results in a reduction of charge on all circuits. There is also a concurrent drop of reserve water in the common stomach, which decreases the number of pulp foragers, which need water to collect building material. New equilibria are then established, accommodating to the lower electron flow (Fig. 7.7). Increasing the water availability for water foragers, or adding water to the nest, is simulated by increasing the output of G with a voltage of plus 0.5 V. This results in a decrease in the charge on the water foragers,

Fig. 7.7 Dynamics of workforce and the common stomach (charges of the circuits). After an equilibrium is found ($t = 2000$ s), a simulated partial removal of the water foragers was initiated. Common stomach: dashed line, builders: thin black line, water foragers: thick gray line, pulp foragers: thick black line. Reprinted from Hilbun and Karsai (2016) with permission from PLOS

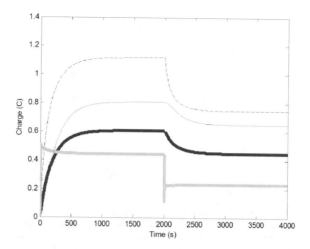

Fig. 7.8 Dynamics of workforce and the common stomach (charges of the circuits). After equilibrium is set on ($t = 2000$ s), extra water was added to the system (i.e., a voltage of 0.5 V was added to the output of G). Common stomach: dashed line, builders: thin black line, water foragers: thick gray line, pulp foragers: thick black line. Reprinted from Hilbun and Karsai (2016) with permission from PLOS

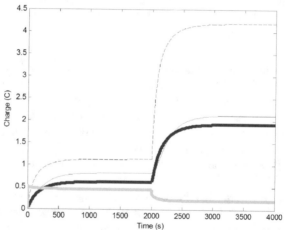

and an increase of the charge of the common stomach. This, in turn, promotes more pulp foraging and building (Fig. 7.8). This kind of behavior is very similar what we can observe in wasp colonies after a rain. Increased water availability will make the refilling of the common stomach easy, so part of the water foragers is converted to pulp foragers and builders.

Based on museum collections and observations, we have analyzed several life history parameters of several wasp species and concluded that variances in the life history, such as the degree of how flexible the behavioral repertoire of the individuals are and how strongly the subsystems are connected via interactions, are correlated (Karsai and Wenzel 1998). The two extremes of this scale are the small societies with independently acting "jack of all trades" individuals and the strongly connected, more rigid behaviorally or age-based caste systems of the large colonies.

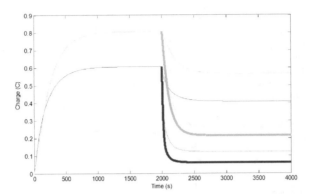

Fig. 7.9 Change of charge in time with different coupling intensity between the circuits ($M1$ and $M2$ were reduced from $0.1H$ to $0.01H$ [thickest line], from $0.1H$ to $0.05H$ (dot/dashed line), and from $0.1H$ to $0.75H$ (medium thick line); pulp foragers (black), and builders (gray). With decreased coupling, the charge on the circuits drops due to a decreased current passed through each circuit by the inductors. Reprinted from Hilbun and Karsai (2016) with a permission from PLOS

Our model, via manipulating the mutual inductance terms and the storage capacity of the common stomach, allowed us to investigate the effect of these interactions quantitatively. Weakening the connections among the elements of the model caused a large decrease of construction related activities, including a drop of water content in the common stomach (Fig. 7.9). This means that the common stomach regulation system is not efficient if the task groups are weakly connected. These societies thus apply the "jack of all trades" workers (Karsai and Wenzel 1998).

It is clear that the two kinds of foragers are connecting to the common stomach differently. Pulp foragers generally spend more time there and a higher number of interactions occurs with common stomach wasps than in the case of water foragers (Karsai and Wenzel 2000). This difference is also reflected in our analysis of the model. The phase-space plots show the existence of point attractors in both cases, but the trajectories are different in nature. Water foragers clearly orbit around the common point attractor and the vicinity of the point attractor is reached very quickly, then the system spirals somewhat away from it (Fig. 7.10) left). The phase-space plot of pulp foragers reveals, by contrast, a consistent dissipation towards the point attractor (Fig. 7.10) right). This suggests that the common stomach's relationship to the water foragers involves a higher degree of negative feedback than the relationship between the common stomach and the pulp foragers.

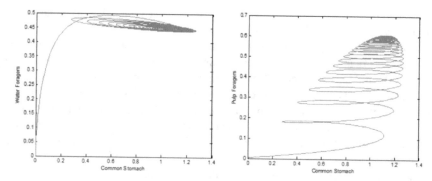

Fig. 7.10 Phase plots of charge of foragers vs. the common stomach. Left panel: Common stomach vs. water foragers with a low common stomach resistance ($R1$ reduced from 5 to 0.000001 Ω). Right panel: Common stomach vs. pulp foragers with a low common stomach resistance ($R2$ reduced from 5 to 0.000001 Ω), The starting point is at the origin. Reprinted from Hilbun and Karsai (2016) with a permission from PLOS

7.6   The Importance and Evolutionary Significance of Common Stomach Regulation

Analyzing the regulation network of task allocation in three insect societies led us from field work through top-down and bottom-up models to exploring circuit models with master equations from physics.

This multitude of approaches helped us recognize that there is a common core of the common stomach regulation system and we concluded that it is more closely akin to the integral control regulation already found in cellular and subcellular levels in biology. The regulatory role of integral control has been predicted, but has not been found on higher organizational levels before (Ang and McMillen 2013; Somvanshi et al. 2015).

This regulatory core, which we call common stomach regulation, describes close regulatory relationships between a substance and the colony members that process that substance. In some sense, the substance itself is regulating its own collection and use by adjusting through a network of feedbacks the workforce that handles this substance. Common stomach regulation is based on simple rules and local interactions, thus it is highly scalable. Scalability is crucial, because the colony size of a given species can range from a few individuals to the millions (Hölldobler and Wilson 2008). This system also needs to be reliable at all scales and this is ensured by a high level of redundancy. The core of the common stomach regulation system is built by a set of 13 negative feedback loops and six positive feedback loops. The insect societies also have some extra elements and feedback loops that are connecting to this core.

Redundancy not only ensures reliability, but also makes the system more resistant to perturbations. The common stomach saturation variable interacts at five different points in the system, and our analyses show that interacting only with the substance outflow and recruitment on one of the task groups would be sufficient for the system to work. While this minimum configuration network is also able to compensate perturbations, it would be less efficient (meaning slower reactions, more oscillations) than the fully connected network. Wasps seem to have a bit simpler network than do bees or ants. They are missing a feedback link from the common stomach to its own outflow, for example. Wasp societies are more "primitive," generally smaller and less homeostatic (i.e., more fluctuations can be observed in their life). Their common stomach contains water, which is probably a low-cost resource compared to prey or pollen protein found as the substance of common stomach in ants and bees.

In the literature, the core of the regulation mechanism of insect societies has commonly been centered on fixed procedures (programs) or on threshold-based mechanisms (reviewed extensively in refs. (Hölldobler and Wilson 2008; Beshers and Fewell 2001)). These models are essentially based on sigmoid response curves paired with a positive feedback control. They are capable of simulating the emergence of division of labor from a principally homogeneous workforce via enlarging small differences in the worker propensity of individuals. While these models have a success to predict experimental data, their modeling assumptions are sometimes less solid. The physical nature of the stimulus is often not specified, and is described instead by a global-state variable of the colony, such as "colony needs" or "food needed for the brood." The mechanisms of how these stimuli and needs are perceived and assessed is often lacking. In several studies, the stimuli for task allocation were identified or predicted as a substance (Karsai and Pénzes 2000; Hölldobler and Wilson 1990; Camazine et al. 1998). We predict that many of the "operative" state variables describing such stimuli (such as colony needs) will prove to be substances that can provide honest signals, due to physical conservation laws. The pure information-based exchange such as "signaling," "messaging," (for example, the bees' dance language) are more vulnerable to outdated or wrong information and to cheating than are substance-based measurements. In fact, the sigmoid stimulus curves can be derived as a by-product of the common stomach regulation mechanism (Agrawal and Karsai 2016). We have also pointed out that a model based on just sigmoid response curves failed a sensitivity test, while the common stomach model of the same phenomena predicted feasible results, even in the case of quite strong perturbations (Schmickl and Karsai 2014).

Besides its role as an information center, the common stomach also provides a strong material buffer against fluctuations. It is acting as an integrator and therefore reduces the impact of system noise (Peng et al. 2009). Resilience against noise decreases the need of a lot of task switching, which could be costly in time (Hamann et al. 2013). As (Ma et al. 2009) emphasized, for three-node networks, having a negative feedback with a buffer node, this is of the most robust available configuration. We emphasized that the common stomach as a buffer is in the center of the common stomach regulation system and has several feedback loops go through the buffer node. The redundancy of the feedback network provides resilience and adaptability.

Robust systems have, in general, the important advantage of having a larger parameter space in which they can evolve and adjust to environmental changes in the long term. This adaptive ability arises primarily from the network connectivity, without requiring the fine-tuning of parameters (Berg and Tedesco 1975; Khan et al. 1995). Identifying and understanding the nature of control mechanisms are of fundamental importance for the understanding of biological regulation. The common stomach regulatory network we discovered is an example of closed loop systems that work on the supra-individual level. This scheme is able to explain both the regulation of work and the resilience of different insect societies. These insect societies evolved their eusociality from non-social predecessors independently, but they all have "discovered" the same regulation mechanism. In the three cases we examined, the same functional process has evolved (convergent evolution), but they are expressed in a different physical manifestation, dependent on the given insect societies' biology and ecology and on their specific substances and interaction types.

References

Agrawal D, Karsai I (2016) The mechanisms of water exchange: the regulatory roles of multiple interactions in social wasps. PLoS One 11(1):e0145560

Ang J, McMillen DR (2013) Physical constraints on biological integral control design for homeostasis and sensory adaptation. Biophys J 104:505–515

Åström KJ, Kumar PR (2014) Control: a perspective. Automatica 50:3–43

Bannister JA, Trivedi KS (1983) Task allocation in fault-tolerant distributed systems. Acta Inform 20:261–281

Berg HC, Tedesco PM (1975) Transient response to chemotactic stimuli in E. coli. Proc Natl Acad Sci USA 72:3235–3239

Beshers SN, Fewell JH (2001) Models of division of labor in social insects. Annu Rev Entomol 46:413–440

Camazine S, Crailsheim K, Hrassnigg N, Robinson GE, Leonhard B, Kropiunigg H (1998) Protein trophallaxis and the regulation of pollen foraging by honey bees (Apis mellifera L.). Apidologie 29:113–126

Crall JD, Gravish N, Mountcastle AM, Kocher SD, Oppenheimer RL, Pierce NE, Combes SA (2018) Spatial fidelity of workers predicts collective response to disturbance in a social insect. Nat Commun 9:1201

Csete ME, Doyle JC (2002) Reverse engineering of biological complexity. Science 295:1664–1669

Del Vecchio D, Dy AJ, Qian Y (2016) Control theory meets synthetic biology. J R Soc Interface 13:20160380

Deneubourg JL, Goss S (1989) Collective patterns and decision-making. Ethol Ecol Evol 1(4):295–311

Deneubourg JL, Pasteels JM, Verhaeghe JC (1983) Probabilistic behaviour in ants: a strategy of errors? J Theor Biol 105:259–271. https://doi.org/10.1016/S0022-5193(83)80007-1

Dussutour A, Beekman M, Nicolis SC, Meyer B (2009) Noise improves collective decision-making by ants in dynamic environments. P R Soc B Biol Sci 276:4353–4361. https://doi.org/10.1098/rspb.2009.1235

Eberlein RL, Peterson DW (1992) Understanding models with Vensim™. In: Morecroft JDW, Sterman JD (eds) Modelling for learning. Eur J Oper Res 59:21–219

Gerkey BP, Matari'c MJ (2004) A formal analysis and taxonomy of task allocation in multi-robot systems. Int J Robot Res 23(9):939–954

Gordon DM (2016) The evolution of the algorithms for collective behavior. Cell Syst. 3:514–520

Hamann H, Karsai I, Schmickl T (2013) Time delay implies cost on task switching: a model to investigate the efficiency of task partitioning. Bull Math Biol 75:1181–1206

Hilbun A, Karsai I (2016) Task allocation of wasps governed by common stomach: a model based on electric circuits. PLoS One 11:e0167041

Hodgkin AL, Huxley A (1952) A quantitative description of membrane current and its application to conduction and excitation in nerve. J Physiol 116:449–556.

Hölldobler B, Wilson EO (1990) The ants. Harvard University Press, Cambridge

Hölldobler B, Wilson EO (2008) The superorganism: the beauty, elegance, and strangeness of insect societies. WW Norton, New York

Huang MH, Seeley TD (2003) Multiple unloadings by nectar foragers in honey bees: a matter of information improvement or crop fullness? Insect Soc 50:330–339. https://doi.org/10.1007/s00040-003-0682-4

Karsai I, Pénzes Z (2000) Optimality of cell arrangement and rules of thumb of cell initiation in *Polistes dominulus*: a modeling approach. Behav Ecol 11:387–395

Karsai I, Phillips MD (2012) Regulation of task differentiation in wasp societies: a bottom-up model of the "common stomach". J Theor Biol 294:98–113

Karsai I, Schmickl T (2011) Regulation of task partitioning by a "common stomach": a model of nest construction in social wasps. Behav Ecol 22:819–830

Karsai I, Wenzel JW (1998) Productivity, individual-level and colony-level flexibility, and organization of work as consequences of colony size. Proc Natl Acad Sci USA 95:8665–8669

Karsai I, Wenzel JW (2000) Organization and regulation of nest construction behavior in Metapolybia wasps. J Insect Behav 13:111–140

Khan S, Spudich JL, McCray JA, Tentham DR (1995) Chemotactic signal integration in bacteria. Proc Natl Acad Sci USA 92:9757–9761

Klügl F, Triebig C, Dornhaus A (2003) Studying task allocation mechanisms of social insects for engineering multi-agent systems. In: Second international workshop on the mathematics and algorithms of social Insects, Atlanta

Koshland DE, Goldbeter A, Stock JB (1982) Amplification and adaptation in regulatory and sensory systems. Science 217:220–225

Li L, Peng H, Kurths J, Yang Y, Schellnhuber HJ (2014) Chaos/order transition in foraging behavior of ants Proc Natl Acad Sci USA 111:8392–8397. https://doi.org/10.1073/pnas.1407083111

Ma W, Trusina A, El-Samad H, Lim WA, Tang C (2009) Defining network topologies that can achieve biochemical adaptation. Cell 138:760–773

O'Donnell S, Bulova SJ (2007) Worker connectivity: a review of the design of worker communication systems and their effects on task performance in insect societies. Insectes Soc 54:203–210

Peng H, Li L, Yang Y, Wang C (2009) Parameter estimation of nonlinear dynamical systems based on integrator theory. Chaos 19:033130. https://doi.org/10.1063/1.3216850

Schmickl T, Karsai I (2014) Sting, carry and stock: How corpse availability can regulate decentralized task allocation in a Ponerine ant colony. PLoS One 9(12):e114611

Schmickl T, Karsai I (2016) How regulation based on a common stomach leads to economic optimization of honeybee foraging? J Theor Biol 389:274–286

Schmickl T, Karsai I (2017) Resilience of honeybee colonies via common stomach: a model of self-regulation of foraging. PLoS One 12(11):e0188004

Schmickl T, Karsai I (2018) Integral feedback control is at the core of task allocation and resilience of insect societies. Proc Nat Acad Sci 115(52):13180–13185. https://doi.org/10.1073/pnas.1807684115

Schmickl T, Thenius R, Crailsheim K (2012) Swarm-intelligent foraging in honeybees: benefits and costs of task-partitioning and environmental fluctuations. Neural Comput Appl 21:251–268

Seeley TD (1983) Division of labor between scouts and recruits in honeybee foraging. Behav Ecol Sociobiol 12:253–259

Seeley TD, Camazine S, Sneyd J (1991) Collective decision-making in honey bees: how colonies choose among nectar sources. Behav Ecol Sociobiol 28(4):277–290

Somvanshi PR, Patel AK, Bhartiya S, Venkatesh KV (2015) Implementation of integral feedback control in biological systems. WIREs Syst Biol Med 7:301–316.

Wiener N (1948) Cybernetics: or control and communication in the animal and the machine. MIT Press, Cambridge

Yi T-M, Huang Y, Simon MI, Doyle J (2000) Robust perfect adaptation in bacterial chemotaxis through integral feedback control. Proc Natl Acad Sci USA 97:4649–4653

Appendix A
Modeling Techniques

A.1 Population Growth Models Using Differential Equations

Our main goal here is to introduce a few modeling techniques we use throughout this book. We do not intend however to provide here the fundamentals on modeling, a tutorial or a review. For these, we refer to other sources (DeAngelis et al. 1992; Ford 2009; Grimm et al. 2006; Kuang 1993). This Appendix is rather a refresher as well as an example of why using different modeling techniques for one and the same problem can be beneficial to understand biological processes better. We start with the simple exponential population growth to make modeling accessible even to complete beginners.

Biologists generally define a population as a collection of individuals that belong to the same species and can potentially breed with each other. One of the best-known early models on population growth was outlined by Malthus (1798). He famously maintained that the human population is predicted to grow in an exponential manner, but the crucial products needed to sustain the population grow in but a linear manner. He argued that these different types of growths will trigger disasters when the population's needs are not satisfied. The basic exponential growth model consists of a single positive feedback loop that arises from the fact that every individual (N) is predicted to have a fixed number of offspring (r), regardless of the size of the population, and thus also regardless of the remaining resources in the habitat:

$$\frac{dN}{dt} = rN \tag{A.1.1}$$

This exponential growth model had a profound effect on biology such as developing the theory of natural selection (Darwin 1859). Exponential population growth is physically not possible in the long run, because either the resources needed to sustain the growth run out (e.g., space, nutrients, access to light, etc.) or due to density-dependent factors, such as the rate of individuals' access to food,

© Springer Nature Switzerland AG 2020
I. Karsai et al., *Resilience and Stability of Ecological and Social Systems*,
https://doi.org/10.1007/978-3-030-54560-4

the dependence of infection on population density, or the density dependence of fertility rates. These will have an increasing effect as the population grows in size. In real populations thus also a negative feedback loop is found within the system that counteracts the positive feedback loop which drives exponential growth. The population usually avoids catastrophe via moderating its own growth as the population increases and the interplay of these two feedback loops also makes the population resilient against perturbations. This general pattern has been observed also in chemistry and it was described by the Belgian mathematician Verhulst (1845) as the density-dependent dynamics of chemical reactions, which led to the best-known early population model:

$$\frac{dN}{dt} = rN\left(1 - \frac{N}{K}\right) \tag{A.1.2}$$

The beauty of this simple equation is that the actual reproduction rate scales linearly between a maximum value (r) and minimum value of zero—or, if a population becomes by some force larger than its habitat's carrying capacity (K), it can even have negative values leading to a decline of the population. No reproduction happens any more when the population reaches its carrying capacity $(N = K)$. This is a simple equation that can be solved formally to predict population size at any time in a closed form, if the maximum reproductive rate (r), carrying capacity (K), and the initial population size (N_0) are given:

$$N(t) = \frac{K N_0 e^{rt}}{K + N_0(e^{rt} - 1)} \tag{A.1.3}$$

Many models in biology have further complexities such as time delays or a large number of other parameters, therefore the actual model equations often cannot be solved formally. In such cases, we can still perform numerical simulations which, with some caveat, will give us still useful predictions and insights into the properties of the system. Many programming languages and modeling platforms have built-in equation solvers. One of the best-known examples for numerical simulations was developed by the school of "System Dynamics" (Forrester 1968) which led to simulation tools like Stella or Vensim (Eberlein and Peterson 1992). The advantage of the systems dynamics approach is that it combines visual sketches that demonstrate the causal relations between the system's components with the mathematical functions needed to solve models (Fig. A.2).

Vensim requires the user to clearly establish the logical connections of the model (by doing sketching). The complicated differential equations are dissected into inflows and outflows, which make debugging easier and only require elemental algebra to plug in the rates. A Vensim system can show results in tables or figures (Fig. A.1).

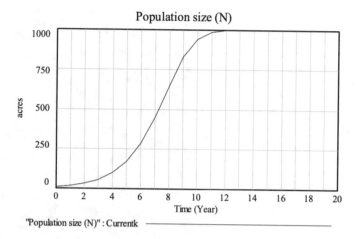

Fig. A.1 Predicted population size per unit area (acres, using the scale given) for a plant growing as the Verhulst equation describes it in Vensim. Parameters: $N_0 = 10$, $r = 0.8$ and $K = 1000$

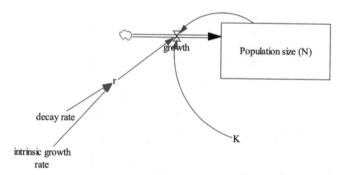

Fig. A.2 Vensim model of sigmoid growth as in the Verhulst equation. The box called "stock" where the quantities (in this case the individuals belonging to the population) are stored. This stock integrates the flows (double arrow). This model has one inflow term called "growth" which funnel the individuals from the source (small cloud) to the stock. This flow contains a rate equation which have 3 inputs and is expressed as an algebraic term: $rN(K - N/K)$

A.2   Agent-Based Models of Population Growth

Without analyzing the sigmoid growth model and its biological importance in details, we also need to mention that the cited assumptions of r and K are biologically quite unrealistic. They are not constants and they actually summarize several other important life history parameters. Generally, the two most common ways to resolve these problems are either to include more parameters and variables into the model (as we did in Chap. 2) or using an altogether different model design, where these parameters are actually missing in the specification of the model but will emerge in runtime.

Fig. A.3 Starting
configuration of an
agent-based Netlogo model.
Trees and animals have their
own rule sets and several
parameters of the model are
actually variables and their
values are different for each
individual. Agents can move
and interact with each other
and with the environment
while updating key variables
such as energy level. This
energy level in turn, affects
the agents' behavior (die of
hunger, breed and so on)

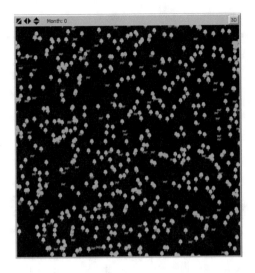

Models using aggregated parameters and equations have a top-down design.
A different, bottom-up design for modeling populating growth would be to use
cybernetic models (urn or cellular automata models, agent-based i.e., individual-
based models) such as implemented in Netlogo (Wilensky 1999). For example, in
an agent-based model for population growth, each agent has an energy level which
is a state of the current energy budget calculated from spending and gaining energy
tokens by the behavior of that individual. Movement and reproduction cost energy,
while feeding gains energy. In this approach, reproduction rate will not be a constant
parameter, but depend, as a simple emergent property, on this energy dynamics. The
same is true for carrying capacity: It is not programmed directly into the system,
but will depend on factors such as how quickly the agents find food, how much they
can exploit food that is present, and how the food regenerates. We show in details
such energy-based ecological microcosms (Chaps. 2 and 3), but the main idea for
these agent- or individual-based models is to simulate population growth where the
aggregate (biologically unrealistic) parameters are replaced by a simple life history
(i.e., a mechanism) with dynamic elements. The agent-based models can also predict
a sigmoid growth without directly implementing r and K (Fig. A.3). Rather, K and
r will be emergent properties of the system (Figs. A.4 and A.5).

While the above two modeling approaches have very different structures,
assumptions, and presentations, their predictions are very close. (The agent-based
model can be downloaded[1]).

To summarize, in this book we wanted to emphasize that different modeling
approaches of the same problems can give the same but also different insights and
understandings. It is important to have different perspectives. Practical questions
also play an important role. For example, agent-based models can run for a long

---

[1] https://sites.google.com/site/springerbook2020/chapter-4.

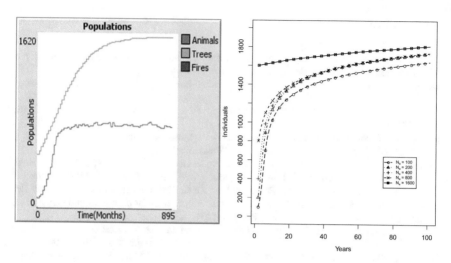

Fig. A.4 Sigmoid population growth predicted by the agent-based model without using the Verhulst equation and the parameters r and K. Left panel: a single run as shown on the modeling platform; Right panel: processed data: average value of 20 runs of the animal population at different starting values (N_0). Reprinted from Karsai et al. (2016) with permission from Elsevier

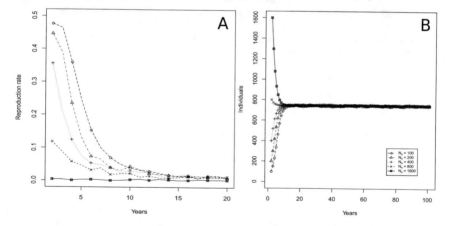

Fig. A.5 Estimated population parameters from the agent-based model (numbers are calculated from the result data and was not part of the model itself); (**A**) Change of reproductive rate of animals; (**B**) emergent stable population size (K) of the animals at different starting conditions. Reprinted from Karsai et al. (2016) with permission from Elsevier

time and needed to have multiple runs and statistical evaluations. This may not be very practical for the cases when quick results are needed. In this case, aggregate models will provide a much faster prediction. On the other hand, agent-based models explore generally a larger state space and provide more variable predictions.

References

Darwin CR (1859) On the origin of species by means of natural selection, or the preservation of favoured races in the struggle for life. John Murray, London

DeAngelis DL, Gross LJ, Boston MA (1992) Individual-based models and approaches in ecology. Chapman and Hall, New York

Eberlein RL, Peterson DW (1992) Understanding models with Vensim™. Eur J Oper Res 59:216–219

Forrester JW (1968) Principles of systems, 2nd edn. Pegasus Communications, Waltham

Ford A (2009) Modeling the environment. Island Press, Washington

Grimm V, Berger U, Bastiansen F, Eliassen S, Ginot V, Giske J, Goss-Custard J, Grand T, Heinz S, Huse G, Huth A, Jepsen JU, Jørgensen C, Mooij WM, Müller B, Pe'er G, Piou C, Railsback SF, Robbins AM, Robbins MM, Rossmanith E, Rüger N, Strand E, Souissi S, Stillman RA, Vabø R, Visser U, DeAngelis DL (2006) A standard protocol for describing individual-based and agent-based models. Ecol Model 198:115–126

Karsai I, Roland B, Kampis G (2016) The effect of fire on an abstract forest ecosystem: an agent based study. Ecol Complexity 28:12–23. https://doi.org/10.1016/j.ecocom.2016.09.001

Kuang Y (1993) Delay differential equations: with applications in population dynamics. Academic Press, Boston

Malthus TR (1798) An essay on the principle of population as it affects the future improvement of society with remarks on the speculations of Mr. Godwin, M. Condorcet, and other writers. J Johnson in St. Paul's Churchyard, London

Verhulst PF (1845) Recherches mathématiques sur la loi d'accroissement de la population. Nouveaux Memoires de l'Academie Royale des Sciences et Belles- Lettres de Bruxelles 18: 1–41.

Wilensky U (1999) NetLogo. Center for connected learning and computer based modeling. Northwestern University, Evanston

Index

© Springer Nature Switzerland AG 2020
I. Karsai et al., *Resilience and Stability of Ecological and Social Systems,*
https://doi.org/10.1007/978-3-030-54560-4

Printed in the United States
by Baker & Taylor Publisher Services